T0290349

Anatomy for the Royal College of Radiologists Fellowship

Illustrated questions and answers

IPEM–IOP Series in Physics and Engineering in Medicine and Biology

About the Series

Series in Physics and Engineering in Medicine and Biology will allow IPEM to enhance its mission to 'advance physics and engineering applied to medicine and biology for the public good.'

Focusing on key areas including, but not limited to:

- clinical engineering
- diagnostic radiology
- informatics and computing
- magnetic resonance imaging
- nuclear medicine
- physiological measurement
- radiation protection
- radiotherapy
- rehabilitation engineering
- ultrasound and non-ionising radiation.

A number of IPEM–IOP titles are published as part of the EUTEMPE Network Series for Medical Physics Experts.

Anatomy for the Royal College of Radiologists Fellowship

Illustrated questions and answers

Andrew G Murchison, Mitchell Chen, Thomas Frederick Barge, Shyamal Saujani, Christopher Sparks, Radoslaw Adam Rippel and Malcolm Sperrin
Oxford University Hospitals, Oxford, UK

Ian Francis
Medical Imaging Partnership, Crawley, UK

IOP Publishing, Bristol, UK

Andrew G Murchison, Mitchell Chen, Thomas Frederick Barge, Shyamal Saujani, Christopher Sparks, Radoslaw Adam Rippel, Malcolm Sperrin and Ian Francis have asserted their right to be identified as the authors of this work in accordance with sections 77 and 78 of the Copyright, Designs and Patents Act 1988.

ISBN   978-0-7503-1185-4 (ebook)
ISBN   978-0-7503-1186-1 (print)
ISBN   978-0-7503-1833-4 (myPrint)
ISBN   978-0-7503-1187-8 (mobi)

DOI   10.1088/978-0-7503-1185-4

Version: 20191201

IOP ebooks

British Library Cataloguing-in-Publication Data: A catalogue record for this book is available from the British Library.

Published by IOP Publishing, wholly owned by The Institute of Physics, London

IOP Publishing, Temple Circus, Temple Way, Bristol, BS1 6HG, UK

US Office: IOP Publishing, Inc., 190 North Independence Mall West, Suite 601, Philadelphia, PA 19106, USA

Contents

Preface

Studying for the anatomy exam is tough but worthwhile, as a good understanding of anatomy forms a key part of the day-to-day life of a radiologist. Whilst you will previously have studied anatomy at medical school or in some cases for surgical exams, you have to approach radiological anatomy in a slightly different way, as it requires you to visualise three-dimensional structures on two-dimensional images. It is also important to be able to recognise structures on the range of imaging modalities that you will encounter in clinical practice, including plain radiographs, ultrasound, fluoroscopy and cross-sectional studies.

We have written this book to try to make the process of learning anatomy as smooth as possible, by providing you with a resource to test your progress and provide advice for revision and sitting the examination. We have written the book specifically with the Royal College of Radiologists Part 1 examination in mind, although the questions and comments are more broadly applicable to other radiology anatomy examinations. We were prompted to write this book because, when we sat the examination ourselves in the past two years, we noticed that some resources featuring exam-style questions were below the level of difficulty that we experienced. We wanted to prepare a text that reflected some of the more difficult questions that you may encounter. Don't be discouraged as you go through the book if you find it particularly challenging–the exam will have a mix of straightforward and difficult questions!

The main difference between our questions and the examination is that the FRCR Part 1 will only have one question per image. We have chosen to test five structures on each image, as we felt this would be a more efficient way to give you as much practice as possible as you work through the book. It also emphasises the point that the structures that you are asked to identify in the exam may be at the periphery of an image, or may be shown from a slightly unusual perspective.

In addition to the questions and answers, we have written comments sections designed to provide further clarification on why we gave certain answers where there might be ambiguity. These sections also describe the anatomy depicted in the questions, to help you to understand how the structures relate to each other and assist with problem-solving in the exam. You will also notice words highlighted in bold–these are structures that we feel could legitimately be examined, and that we would advise you to be able to recognise (although not necessarily on the image provided). The exam tips are taken from our experiences, and include advice we were given during preparation for the Part 1, and reflections from our own experience.

Although we have made every effort to make sure we have been as accurate as possible, the text may still contain mistakes, so please do not rely on this book for clinical decision making. If you do notice any mistakes, we would be grateful to receive your feedback, so that we can improve any future editions of the book. Finally, we would advise that you read the guidance provided on the Royal College of Radiologists website, particularly in case any changes in format or content have arisen since the publication of this book.

Thanks for reading, and good luck!

Author biographies

Andrew G Murchison

Andrew grew up in Edinburgh and studied medicine at the University of Oxford. He was awarded the top first for the BA in medical sciences, and graduated with distinction. He completed core medical training and is currently an academic clinical fellow in clinical radiology. He is married with a young son, and lives in Oxford.

Mitchell Chen

Mitch is an academic clinical fellow in radiology at the Oxford University Hospitals NHS Foundation Trust, where he researches on novel imaging modalities alongside his clinical training. He graduated from Oxford University, with a bachelor's degree in medicine and surgery, a master's degree in engineering science and a DPhil. In addition to his work, he has passions for long distance running and travelling.

Thomas Frederick Barge

Having graduated in medicine from the University of Oxford in 2011, Tom completed his foundation training in the Oxford deanery. He then moved to New Zealand where he worked in acute medicine, ED and ICU, before returning to start radiology training in Oxford. He is a keen cyclist.

Shyamal Saujani

Dr Shyamal Saujani is a third year radiology registrar in the Oxford training programme. He undertook his undergraduate medical training at the Universities of Exeter and Plymouth, before embarking on an intercalated degree in anatomy at King's College University, London. In his spare time he enjoys cooking, spending time with his family and travelling around the country following his favourite football club.

Christopher Sparks

Upon completion of his undergraduate studies and the Foundation Programme in Liverpool, Chris worked in a variety of medical specialties in New Zealand before returning to the UK for Core Medical Training and, subsequently, the Clinical Radiology training programme in Oxford. Outside of work Chris enjoys running, skiing, travelling and the search for the perfect British pub.

Radoslaw Adam Rippel

Radoslaw Rippel is a Clinical Radiology registrar in Oxford Deanery and University College London Medical School alumnus. His previous work includes publications on use of nanoparticles, tissue engineering and endovascular robotics. Clinically his main interests are musculoskeletal radiology and ultrasound guided intervention. In his spare time Radoslaw enjoys motorsports, cycling and running. Recently he has welcomed a baby girl however, and so spare time has become a thing of the past.

Malcolm Sperrin

Malcolm was born in Cuba of diplomatic parents in 1963, and attended The Harvey Grammar School in Folkestone leaving there in 1981 to study Physics with Maths at Reading University. His first job was working on Artificial Intelligence and then with the UK Atomic Energy Authority on reactor fault analysis. This experience placed him in a good position to provide insight into both the Chernobyl and Fukushima incidents.

After further study at Reading University, Malcolm joined Medical Physics at the Churchill Hospital in Oxford with responsibility for non-ionising radiation. In 1995, Malcolm moved to the Princess Margaret Hospital in Swindon acting as Deputy Head of Department and then, in 2002, he moved to The Royal Berkshire Hospital in Reading taking on the role of Departmental Director.

Malcolm has a special interest in radiation medicine, especially Nuclear Medicine and Radiotherapy. He also plays a significant role in radiation protection and contingency planning. In parallel to his conventional hospital duties, Malcolm also spends a lot of time teaching and lecturing with organisations including Oxford Postgraduate Medical School, The Open University and various Royal Colleges not to mention lectureships at Guildford and the University of the West of England.

Malcolm was made a visiting professor at Reading, Guildford and Open Universities and visiting academic at Oxford University and plays a role on the national stage with the Institute of Physics, Royal Institution, Science Media Centre and the British Association for the Advancement of Science. Malcolm also feeds into activities centred on science and health policy at the DoH.

Malcolm's down-to-earth approach to Medical Science has led to him being frequently sought by the media for comment on mobile phone use, WiFi safety and even the risks from the Fukushima reactor. He is very active in developing innovation whether operational or scientific and has recently been involved in initiatives with Microsoft and other multi-national companies with a drive to improve patient outcomes.

Malcolm is a keen adventure sports enthusiast and likes to climb, cave and canoe and has been known to parachute. He has a partner, Nicki (who is not sure about the parachuting), an 9-year-old son and a spaniel called Harvey.

Ian Francis

Dr Ian Francis MA (Clin Ed), FRCR, FRCS, BDS (Hons) is a consultant radiologist working across both the NHS and private sectors. He is also a co-founder and a director at Medical Imaging Partnership, a clinico-radiological business supporting radiologists and imaging departments in developing strategy and service redesign of their imaging services with fully integrated back office services. Ian Is also a leader in educational development across the radiology spectrum and was the Education Development Lead for the Royal College of Radiologists, driving programmes of radiology educational delivery throughout the UK.

IOP Publishing

Anatomy for the Royal College of Radiologists Fellowship

Illustrated questions and answers

Andrew G Murchison, Mitchell Chen, Thomas Frederick Barge, Shyamal Saujani, Christopher Sparks, Radoslaw Adam Rippel, Malcolm Sperrin and Ian Francis

Chapter 1

Head and neck

Andrew G Murchison and Mohammed Khoshkoo

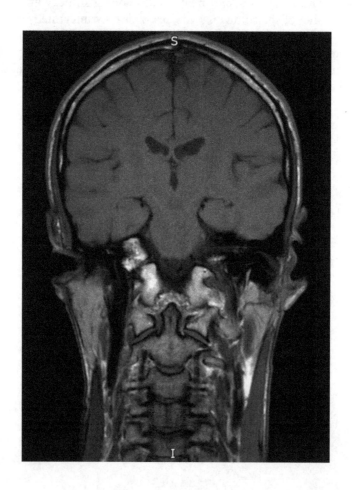

Q1.1 3D reconstruction of a paediatric skull CT

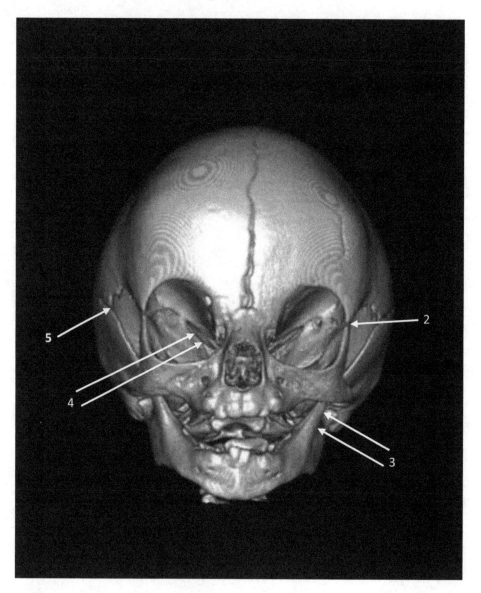

1. Name the anatomical variant.
2. Name the arrowed structure.
3. Name the arrowed structure.
4. Name the arrowed structure.
5. Name the arrowed structure.

Answers
1. Metopic suture.
2. Left frontozygomatic suture.
3. Left coronoid process of the mandible.
4. Right superior orbital fissure.
5. Right pterion.

Comments:
The **frontal bone** of the skull is separated from the two **parietal bones** by the **coronal suture**. The parietal bones meet at the **sagittal suture**, which forms a 'T-junction' with the coronal suture at a point called the **bregma**. The point at which the sagittal suture meets the **lambdoid suture**, separating the parietal bones from the **occipital bone**, is known as the **lambda**.

The **temporal bones** join the parietal bones at the **squamous sutures**. The junction where the coronal and squamous sutures, and the frontal, parietal, temporal and **sphenoid bones** converge is known as the **pterion**.

Exam tips:
- A metopic suture, which is found most commonly in infants but which may persist into adulthood, is an exam favourite. If the question asks for an anatomical variant, and the image presented is of the skull, there's a good chance that you'll find a metopic suture.
- Remember to be specific if the arrows are clearly pointing to a specific part of a larger structure (e.g. the coronoid process of the mandible).

Q1.2 Lateral radiograph of the facial bones of a child

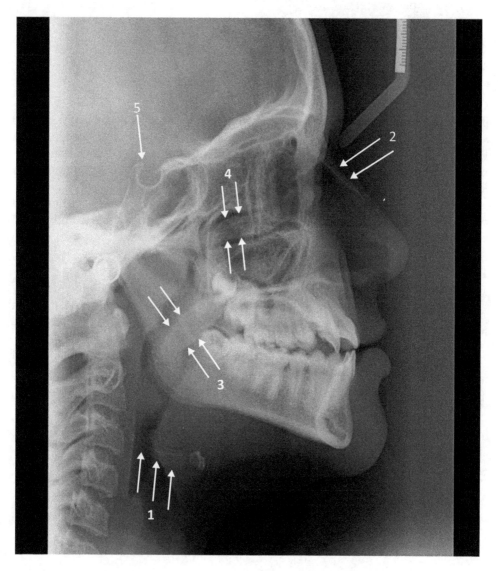

1. Name the arrowed structure.
2. Name the arrowed structure.
3. Name the arrowed structure.
4. Name the arrowed structure.
5. Name the arrowed structure.

Answers
1. Hyoid bone.
2. Nasal bone.
3. Soft palate.
4. Middle nasal turbinate.
5. Posterior clinoid process.

Comments:
The **nasal turbinates** (or **conchae**) are ledges of bone which arise from the lateral walls of the nasal cavity. The **inferior meatus** is the nasal passage below the inferior turbinate, the **middle meatus** lies between the inferior and middle turbinates, and the **superior meatus** lies between the superior and middle turbinates.

The **hard palate** is a bony structure which forms the superior border of the oral cavity and the inferior border of the nasal cavity. Posterior to this is the **soft palate**, a soft tissue structure containing several muscles involved in swallowing. Projecting from the posterior border of the soft palate is protuberance called the **uvula**.

Exam tip:
- The inferior nasal turbinate can be readily identified because it is the largest of the turbinates.

Q1.3 Coronal bony reconstruction from a CT petrous bones study

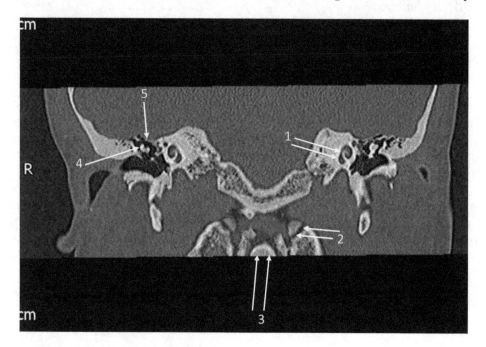

1. Name the arrowed structure.
2. Name the arrowed structure.
3. Name the arrowed structure.
4. Name the arrowed structure.
5. Name the arrowed structure.

Answers

1. Left cochlea.
2. Left atlantooccipital joint.
3. Odontoid process of C2 (dens).
4. Right head of malleus.
5. Right tegmen tympani.

Comments:

The inner ear and structure of the petrous part of the temporal bone are complex but could legitimately pop up in the exam.

The **external auditory canal** leads to the tympanic membrane—this is tethered at its superior margin to a bony promontory called the **scutum** (marked with a red asterisk). Within the middle ear are the ossicles—the **malleus, incus** and **stapes**. The base of the stapes attaches to the oval window of the inner ear, which comprises the shell-shaped **cochlea** and three **semi-circular canals**.

The roof of the middle ear is separated from the meninges and brain by a thin ceiling of bone called the **tegmen tympani**. A canal called the **additus ad antrum** opens from the superior middle ear into an air chamber within the petrous bone called the **maxillary antrum**.

Several important structures run through the petrous bone. The facial nerve enters the skull base through the **internal auditory meatus**. It travels through the petrous part of the temporal bone in the **facial canal,** which has a question-mark-shaped course, before leaving the skull at the **stylomastoid foramen**. The **internal carotid artery** also passes through the petrous temporal bone in the **carotid canal**.

Exam tips:

- The incus and malleus form an ice-cream cone configuration—with the head of the malleus forming the ice cream, and the body of the incus forming the cone.
- Unusual views of common structures are an exam favourite. If (when!) you find that you have no idea what the arrow's pointing at, try to orientate yourself by identifying surrounding structures. For example, for question 2, you could identify the odontoid process of C2 poking up in the midline from the bottom of the image. That would make the two bony structures on either side the lateral masses of C1—which articulate superiorly with the skull at the atlanto-occiptal joints.

Q1.4 Axial slice from a CT cisternogram study with intrathecal contrast

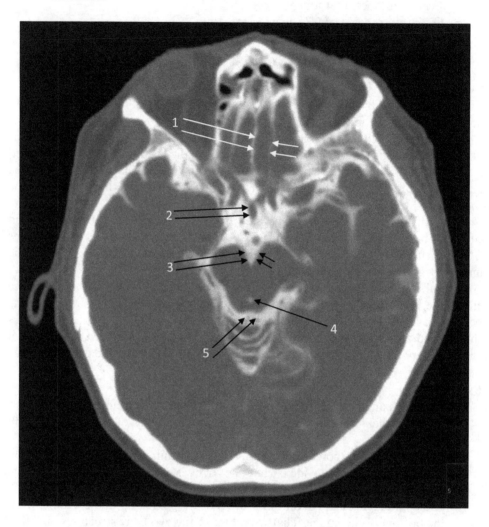

1. Name the arrowed structure.
2. Name the arrowed structure.
3. Name the arrowed structure.
4. Name the arrowed structure.
5. Name the arrowed structure.

Answers
1. Left gyrus rectus (straight gyrus).
2. Pituitary infundibulum (pituitary stalk).
3. Interpeduncular cistern.
4. Cerebral aqueduct (aqueduct of Sylvius).
5. Quadrigeminal cistern.

Comments:
The basal cisterns are a network of cerebrospinal fluid (CSF)—filled spaces which communicate with the ventricular system via the Foramen of Magendie (also called the median or medial aperture) and the two lateral Foramina of Luschka (lateral apertures)—these drain into the **cisterna magna**, which sits at the base of the skull between the **medulla** and **cerebellum**.

Cisterns that you should be able to identify include the **suprasellar cistern**, which sits above the **pituitary fossa** (sella turcica) and contains important structures including the Circle of Willis and the optic chiasm. The suprasellar cistern has a star-shaped configuration, with the posterior point of the star formed by the **interpeduncular fossa**, which lies between the two **cerebral peduncles** of the midbrain. The two posteriolateral points of the star are the **ambient cisterns** which are lateral to the brain stem; these join posteriorly behind the midbrain to form the **quadrigeminal cistern**. The anterolateral points of the star are the **Sylvian fissures** (lateral sulci), which separate the frontal and parietal lobes from the temporal lobes, and contain the **middle cerebral arteries**.

Other cisterns to recognise are the **prepontine** and **premedullary** cisterns, which are best seen on sagittal images and lie in front of the **pons** (with its' belly) and the medulla, respectively.

Exam tips:
- The quadrigeminal cistern can be mistaken for the fourth ventricle. The quadrigeminal cistern is seen at the level of the midbrain (which has 'Mickey Mouse ears')—the fourth ventricle is seen at the level of the pons. An additional clue in question 5 is that the cerebral aqueduct, which drains from the third to the fourth ventricle, is also visualised; therefore this cannot be the fourth ventricle.
- Question 2 is difficult because it is an obscure view of a common structure. The key here is to orientate yourself—it is a midline structure in the suprasellar cistern, lying just behind the optic chiasm.

Q1.5 Midline sagittal section from a CT venogram with intravenous contrast

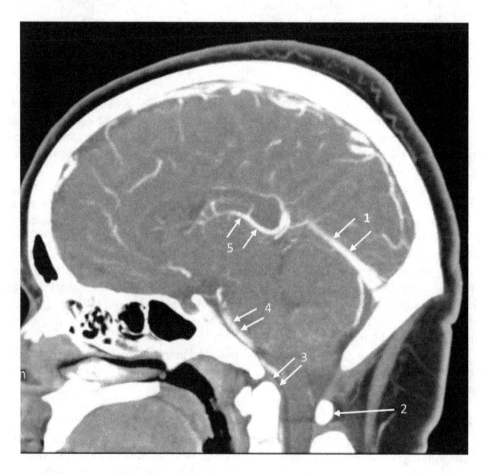

1. Name the arrowed structure.
2. Name the arrowed structure.
3. Name the arrowed structure.
4. Name the arrowed structure.
5. Name the arrowed structure.

Answers
1. Straight sinus.
2. Posterior arch of C1 (atlas).
3. Tectorial membrane.
4. Basilar artery.
5. Internal cerebral vein.

Comments:
The **superior sagittal sinus** meets the **right** and **left transverse sinuses** at the **confluence of the sinuses (torcular herophili)**. The transverse sinuses become the **sigmoid sinuses**, which leave the skull base through the **jugular foramen** to form the **internal jugular veins**. The sigmoid sinus becomes the internal jugular vein when it is joined by the inferior petrosal sinus in the jugular foramen.

The right and left **basal veins of Rosenthal** join with the **internal cerebral veins** to form the **vein of Galen**, which is short and runs within the quadrigeminal cistern. This then combines with the **inferior sagittal sinus** to form the **straight sinus**, which drains into the confluence of the sinuses.

The **tectorial membrane** is continuous with the **posterior longitudinal ligament** of the spine, and passes from the posterior body of C2, along the posterior **odontoid process**, to the **clivus**.

Exam tips:
- Remember that even though the phase of contrast is venous, arterial structures (such as the basilar artery in this case) may still be discernible, and vice versa.
- In most cases in the exam, if the structure shown is paired, then you should include left or right in your answer. In some cases, however, it isn't possible to tell (as with the internal cerebral vein here)—in that case, just give the name of the structure without including laterality.

Q1.6 Axial slice from a T2-weighted MRI of the brain

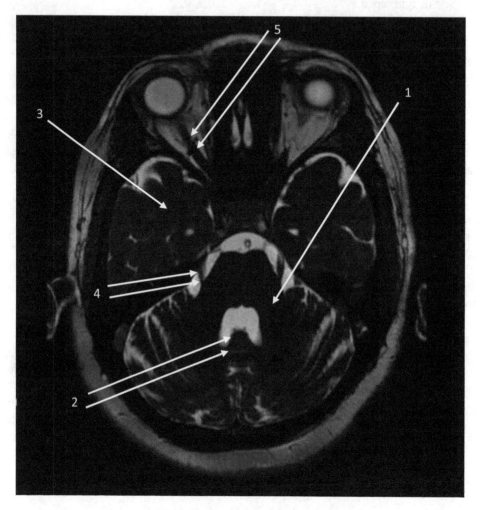

1. Name the arrowed structure.
2. Name the arrowed structure.
3. Name the arrowed structure.
4. Name the arrowed structure.
5. Name the arrowed structure.

Answers
1. Left middle cerebellar peduncle.
2. Cerebellar vermis.
3. Right temporal lobe.
4. Right trigeminal nerve.
5. Right optic nerve.

Comments:
The main cranial nerves that are identifiable on cross-sectional imaging and are therefore most likely to appear in the exam are the **optic nerve**, the **trigeminal nerve**, the **facial nerve** and the **vestibulocochlear nerve**. The trigeminal nerve, which provides sensory input from the face and also innervates the muscles of mastication, arises from the anterior pons. It travels anteriorly through the **prepontine cistern** and into the **trigeminal (Meckel's) cave**, where it forms the trigeminal (Gasserian) ganglion. At this point, it divides into three branches. V1 (the ophthalmic nerve) passes through the **cavernous sinus** and then the **superior orbital fissure**. V2 (the maxillary nerve) also passes through the cavernous sinus, and exits the skull through the **foramen rotundum**. V3 (the mandibular nerve) exits the skull through the **foramen ovale**.

The optic nerve arises at the retina, and travels posteriorly through the orbit to enter the skull through the **optic canal**. It then passes through the **suprasellar cistern** to join with the contralateral optic nerve at the **optic chiasm**. Nerve fibres then travel in two **optic tracts** to the lateral geniculate nuclei.

The **cerebellar peduncles** are three paired white matter tracts that connect the brainstem with the cerebellum. The **middle cerebellar peduncles** are the largest of these, and connect the **pons** and cerebellum. The superior and inferior cerebellar peduncles connect the cerebellum with the **midbrain** and **medulla**, respectively.

Exam tip:
- The trigeminal (Meckel's) cave isn't shown on this image, but is a classic exam question—make sure you're familiar with identifying it on an axial MRI section.

Q1.7 Midline sagittal image from a T2-weighted MRI brain

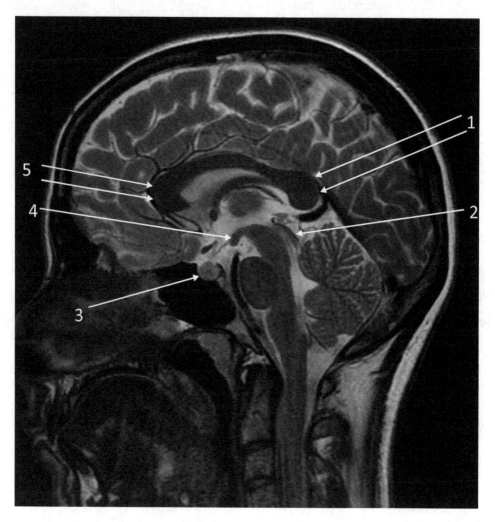

1. Name the arrowed structure.
2. Name the arrowed structure.
3. Name the arrowed structure.
4. Name the arrowed structure.
5. Name the arrowed structure.

Answers
1. Splenium of the corpus callosum.
2. Tectal (quadrigeminal) plate.
3. Pituitary gland.
4. Mamillary body.
5. Genu of the corpus callosum.

Comments:
The midline sagittal section commonly appears in examinations because it demonstrates a lot of anatomy. It is important to recognise the brainstem structures of the **midbrain, pons** and **medulla,** and also the **cerebral aqueduct** and **fourth ventricle**.

The midbrain is divided by the plane of the cerebral aqueduct into the larger tegmentum ventrally and the smaller tectum dorsally. The tectum is often referred to as the **tectal (quadrigeminal) plate**, and is readily identifiable on sagittal images.

The **corpus callosum** links the right and left cerebral hemispheres and is divided into four main parts from anterior to posterior: the **rostrum**, the **genu** (knee), the **body**, and finally the more bulbous **splenium** (meaning bandage). Additionally, the thin section between the body and splenium is sometimes described as the isthmus.

The **thalami** sit on either side of the **third ventricle**. They are connected by a band of tissue called the **massa intermedia** (interthalamic adhesion), although this is not present in all patients.

Exam tip:
- Note that some of the structures in the upper spine are demonstrated here—including the dens and the anterior and posterior arches of C1.

Q1.8 Axial image from T2-weighted sequence of MRI brain

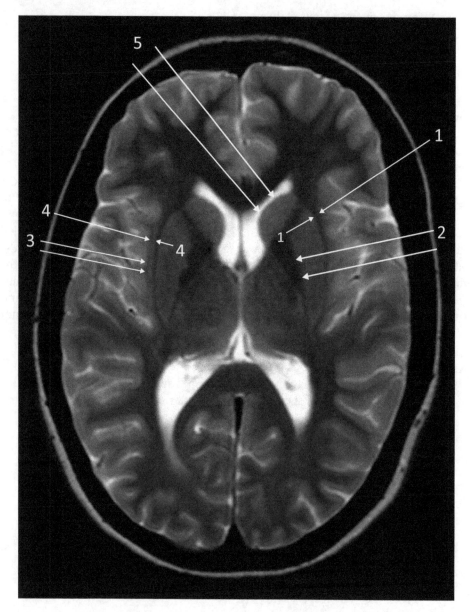

1. Name the arrowed structure.
2. Name the arrowed structure.
3. Name the arrowed structure.
4. Name the arrowed structure.
5. Name the arrowed structure.

Answers
1. Left external capsule.
2. Posterior limb of the left internal capsule.
3. Right extreme capsule.
4. Right claustrum.
5. Head of the left caudate nucleus.

Comments:
The basal ganglia are deep (i.e. not at the cortical surface) grey matter structures. The paired **caudate nuclei** are tucked in beside the lateral ventricles—they are comma-shaped, with the bulbous **head of the caudate nucleus** anteriorly. The other main structure in the basal ganglia is the **lentiform nucleus** comprising the **putamen** and the **globus pallidus** (these are not distinguishable on the image in this question). This is separated from the caudate nucleus by the **internal capsule**. The internal capsule is a white matter structure which is a major avenue of communication between the cortex and the lower CNS; for example, the corticospinal tract passes through it. It is divided into three parts—the **anterior limb** (between the lentiform nucleus and caudate nucleus) and the **posterior limb** (between the lentiform nucleus and **thalamus**), which are connected by the **genu** ('knee').

Beyond the lentiform nucleus is an onion-skin of structures—the **external capsule** (a white matter tract) beyond which is the **claustrum** (a strip of grey matter). Next along is another white matter tract, the **extreme capsule**, followed by the **insular cortex**.

Exam tips:
- The axial view through the basal ganglia is another exam classic. Also make sure to be familiar with the anatomy in coronal views.
- **Cavum septum pellucidum** and **cavum vergae** are often shown on coronal sections—look out for these if the question asks for an anatomical variant.

Q1.9 T1-weighted coronal slice from an MRI in an adult patient

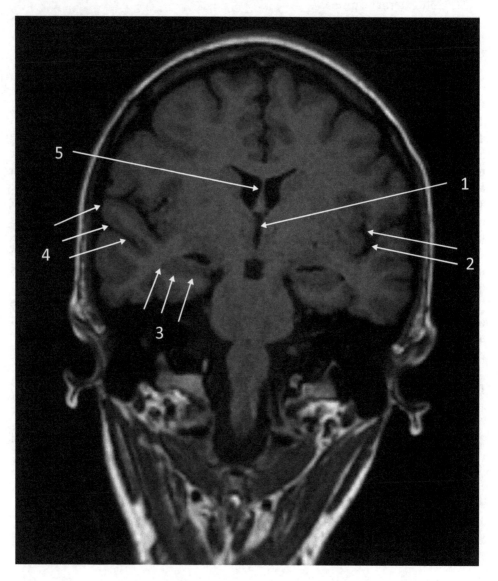

1. Name the arrowed structure.
2. Name the arrowed structure.
3. Name the arrowed structure.
4. Name the arrowed structure.
5. Name the arrowed structure.

Answers
1. Third ventricle.
2. Left lateral sulcus (or lateral fissure/Sylvian fissure).
3. Right hippocampus.
4. Right superior temporal gyrus.
5. Septum pellucidum.

Comments:
The **hippocampus** is a grey matter structure most readily identifiable on coronal images. It lies on the medial side of the **temporal lobe**, immediately inferior to the **temporal horn of the lateral ventricle**. The **amygdala** is anterior to the hippocampus, and curls round to lie anterior and superior to the temporal horn of the lateral ventricle.

The **fornix** is a C-shaped white matter tract that arises posteriorly from the hippocampus. One crux of the fornix arises from each hippocampus, and curls superiorly and anteriorly until it joins with the contralateral crux. These form the **body of the fornix** which lies immediately below the **septum pellucidum**. As it continues anteriorly, the fornix splits again into two columns. Each **column** travels in an inferoposterior arc behind the **anterior commissure** and then lateral to the **third ventricle**. It then meets the ipsilateral **mamillary body**. The mamillary bodies lie at the base of the brain below the third ventricle and are seen anterior to the midbrain on MRI.

Exam tip:
- The distinction between the hippocampus and the amygdala is difficult on a single coronal image. However, the amygdala is located more anteriorly, and we would not expect to see it together with the brainstem on the same coronal image.

Q1.10 Coronal image from a CT angiography study with intravenous contrast

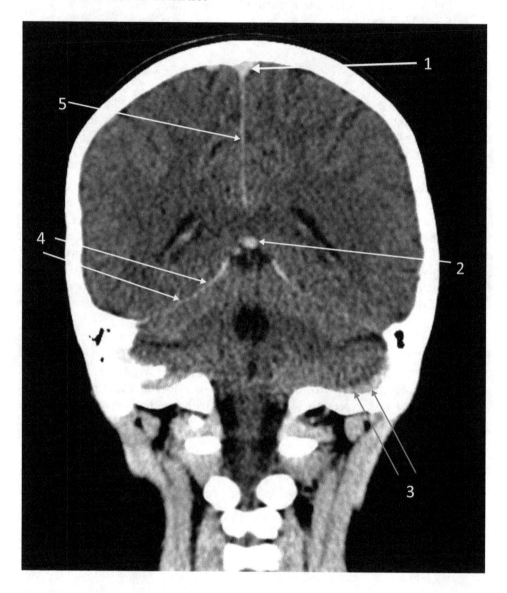

1. Name the arrowed structure.
2. Name the arrowed structure.
3. Name the arrowed structure.
4. Name the arrowed structure.
5. Name the arrowed structure.

Answers
1. Superior sagittal sinus.
2. Straight sinus.
3. Left sigmoid sinus.
4. Tentorium cerebelli.
5. Falx cerebri.

Comments:
The **falx cerebri** is a fold of dura mater which is situated in the **interhemispheric fissure** and divides the two cerebral hemispheres. It attaches to the **crista galli** anteriorly and to the **tentorium cerebelli** posteriorly.

The **tentorium cerebelli** is another fold of the dura mater, which lies above the cerebellar hemispheres. The falx cerebri splays slightly where it meets the tentorium, creating a pouch which contains the **straight sinus**.

Exam tip:
- Remember to read the description at the start of the question as it can give useful clues to identify structures. In this case, the information about contrast administration gives a clue that hyperattenuating structures may be vascular—namely, cerebral venous sinuses. However, bear in mind that other structures might still appear bright even though they aren't vascular—as with the falx cerebri and tentorium in this image.

Q1.11 T1-weighted sagittal section from an MRI of the brain

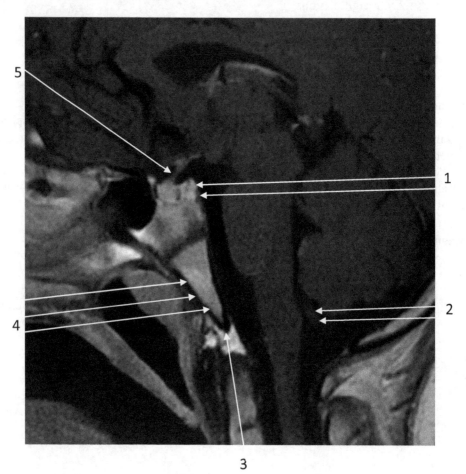

1. Name the arrowed structure.
2. Name the arrowed structure.
3. Name the arrowed structure.
4. Name the arrowed structure.
5. Name the arrowed structure.

Answers
1. Dorsum sellae.
2. Cerebellar tonsil.
3. Basion.
4. Clivus.
5. Pituitary infundibulum.

Comments:
The **sella turcica** (also known as the **hypophyseal or pituitary fossa**) is a hollow in the **sphenoid bone** which contains the **pituitary gland**. The anterior boundary of the pituitary fossa is the **tuberculum sellae**, above which are two small processes called the **anterior clinoid processes**. The posterior boundary of the pituitary fossa is the **dorsum sellae** ('back of the seat'), which is continuous inferiorly with the **clivus**. Two processes at top of the dorsum sellae are known as the **posterior clinoid processes**.

The clivus is a backward-sloping structure in the skull base which is formed by part of the **sphenoid bone** superiorly and part of the **occipital bone** inferiorly. Posterior to the clivus sits the **basilar artery** and the **pons**. The inferior tip of the clivus is an anatomical landmark known as the **basion**—this is the anterior margin of the **foramen magnum**.

Exam tip:
- On a T1-weighted MRI, the posterior part of the pituitary gland (neurohypophysis) is brighter than the anterior pituitary (adenohypothysis).

Q1.12 3D reconstruction of a phase contrast angiography MRI sequence

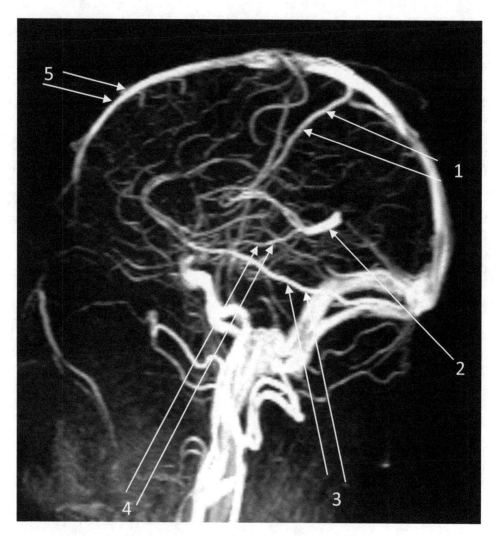

1. Name the arrowed structure.
2. Name the arrowed structure.
3. Name the arrowed structure.
4. Name the arrowed structure.
5. Name the arrowed structure.

Answers
1. Superior anastomotic vein (Vein of Trolard).
2. Vein of Galen.
3. Inferior anastomotic vein (Vein of Labbé).
4. Basal vein (of Rosenthal).
5. Superior sagittal sinus.

Comments:

The **superior anastomotic veins** (Veins of Tolard) connect the **superior sagittal sinus** to the superficial middle cerebral veins, which ultimately drain to the **cavernous sinuses**. The **inferior anastomotic veins** (Veins of Labbé) connect the **transverse sinuses** to the superficial middle cerebral veins.

The right and left **basal veins of Rosenthal** join with the **internal cerebral veins** to form the **vein of Galen**, which is short and runs within the **quadrigeminal cistern**. This then combines with the **inferior sagittal sinus** to form the **straight sinus**, which drains into the **confluence of the sinuses**.

Exam tip:
- This is a difficult question. The superior and inferior anastomotic veins can be identified because they are large veins draining into the superior sagittal sinus and transverse sinus, respectively.

Q1.13 Coronal section from an MRI study of the orbits

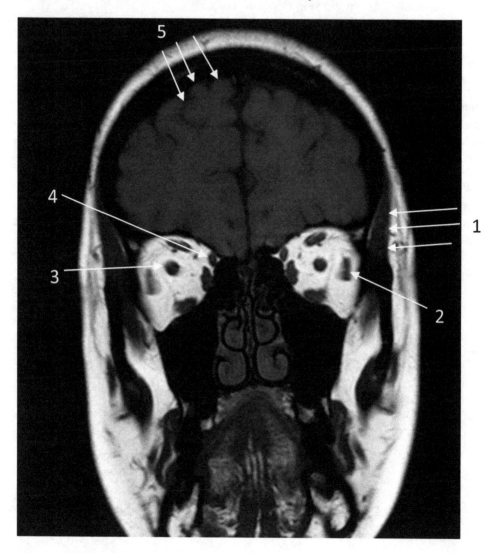

1. Name the arrowed structure.
2. Name the arrowed structure.
3. Name the arrowed structure.
4. Name the arrowed structure.
5. Name the arrowed structure.

Answers
1. Left temporalis muscle.
2. Left lateral rectus muscle.
3. Right optic nerve.
4. Right superior oblique muscle.
5. Right superior frontal gyrus.

Comments:

The four rectus muscles—**superior** and **inferior**, **lateral** and **medial**, are well demonstrated on a coronal MRI, and the lateral and medial rectus muscles are also common questions on axial imaging. The lateral rectus muscle is innervated by the abducens nerve (6th cranial nerve), whilst the other rectus muscles are innervated by the oculomotor nerve (3rd cranial nerve).

The **superior oblique muscle** originates from the medial side of the orbit, superior to the medial rectus. It is innervated by the trochlear nerve (4th cranial nerve). The inferior oblique is a small muscle which, unlike the other extra-ocular muscles, does not have its origin at the orbital apex but instead originates from the maxillary bone in the anterior orbit. Like three of the rectus muscles, it is innervated by the oculomotor nerve.

In the eye itself, the **lens** and the **vitreous humour** behind it are readily demonstrable on CT or MRI.

The **optic nerve** travels posteriorly through the orbit from the retina. It travels through the **optic canal** into the **suprasellar cistern**, where it meets the contralateral nerve at the **optic chiasm**.

Exam tip:
- The **lacrimal gland** can sometimes crop up as a sneaky question on axial or coronal sequences, and is found at the superolateral aspect of the orbit.

Q1.14 Parasagittal T1-weighted sequence from an MRI head

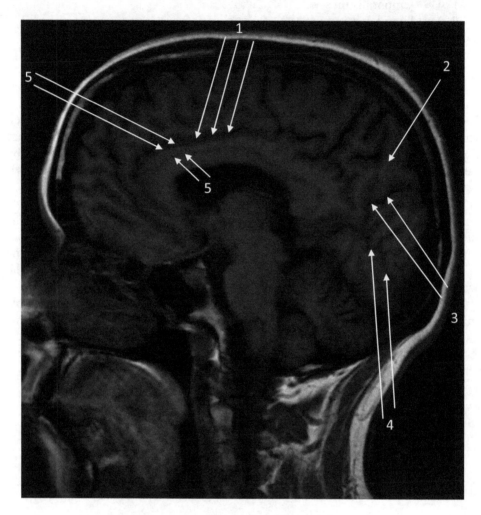

1. Name the arrowed structure.
2. Name the arrowed structure.
3. Name the arrowed structure.
4. Name the arrowed structure.
5. Name the arrowed structure.

Answers
1. Cingulate sulcus.
2. Parietal lobe.
3. Parieto-occipital sulcus.
4. Calcarine sulcus.
5. Cingulate gyrus.

Comments:
There are four lobes within each brain hemisphere. The **frontal lobe** is the largest, and is separated from the **parietal lobe** by the **central sulcus**. The **pre-central gyrus** (motor strip) is part of the frontal lobe immediately anterior to the **central sulcus**, and is responsible for motor control. The **post-central gyrus** is immediately posterior to the central sulcus and contains the primary sensory cortex.

The **parieto-occipital sulcus** lies between the parietal lobes and the **occipital lobes**. The visual cortex is situated on both sides of the **calcarine sulcus**, which sits in the middle of the occipital lobe, and joins with the parieto-occipital sulcus medially.

The **temporal lobes** are located laterally—they are separated from the frontal lobes by the **Sylvian fissure** anteriorly and are continuous with the parietal and occipital lobes posteriorly.

It is also important to recognise the paired **cingulate gyri** (which lie immediately above the corpus callosum) and the **cingulate sulci** superior to them.

Exam tip:
- If possible, be specific when naming parts of the cerebral cortex, e.g. by giving the answer 'superior frontal gyrus' rather than 'frontal lobe' on an appropriate coronal image. However, sometimes it isn't possible to be more specific, as with the parietal lobe question above—in that case, give the name of the lobe.

Q1.15 Axial section from a T2-weighted MRI of the brain

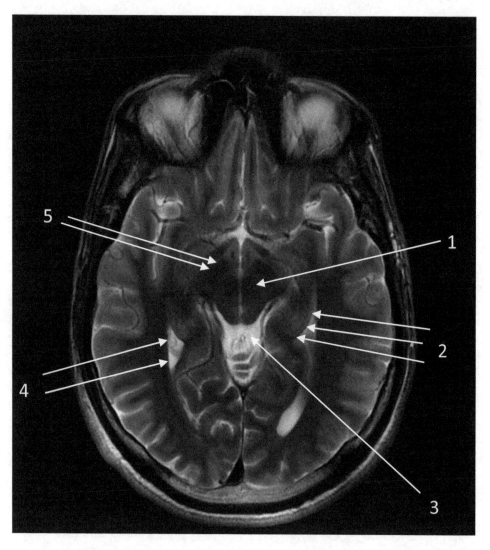

1. Name the arrowed structure.
2. Name the arrowed structure.
3. Name the arrowed structure.
4. Name the arrowed structure.
5. Name the arrowed structure.

Answers
1. Left red nucleus.
2. Left hippocampus.
3. Quadrigeminal cistern.
4. Temporal horn of the right lateral ventricle.
5. Right substantia nigra.

Comments:
Anteriorly within the **midbrain** are the **cerebral peduncles** which connect the brainstem with the higher CNS structures. Just posterior to these on axial slices are bilateral strips containing dopaminergic cell bodies—the **substantia nigra**. The **red nucleus** is a spherical structure of cell bodies that is identifiable on MRI and is located posterior to the substantia nigra at certain axial levels.

The **cerebral aqueduct** runs perpendicularly through the midline of the posterior part of the midbrain, between the **third** and **fourth ventricles**. This is surrounded by a ring of grey matter—the **peri-aqueductal grey**. Behind the coronal plane containing the cerebral aqueduct is the **tectal (quadrigeminal) plate**. On the posterior side of the tectal plate are two sets of rounded protrusions called the **superior** and **inferior colliculi,** beyond which lies the **quadrigeminal cistern**. The superior colliculi are at the same axial level as the red nucleus.

Exam tip:
- The **midbrain** is recognisable on axial brain sections because of its' 'Mickey Mouse' ears anteriorly, formed by the cerebral peduncles.

Q1.16 Angiographic study with contrast injection into the common carotid artery

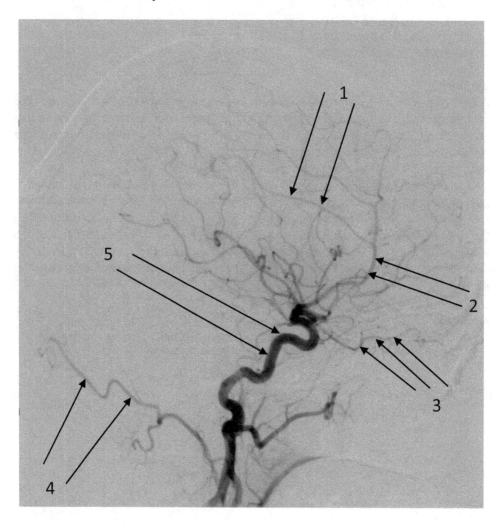

1. Name the arrowed structure.
2. Name the arrowed structure.
3. Which of the cranial foramina does this structure pass through?
4. Name the arrowed structure.
5. Name the arrowed structure.

Answers
1. Pericallosal artery.
2. Anterior cerebral artery.
3. Optic canal.
4. Occipital artery.
5. Internal carotid artery.

Comments:

The **common carotid artery** gives rise to the **internal carotid artery** and **external carotid artery**. The external carotid artery has six main branches and two terminal divisions—the **occipital artery** is a large branch originating from the posterior side.

The internal carotid artery has a characteristic shape on angiography and is divided into seven segments. The **ophthalmic artery** arises distally from the internal carotid artery after it leaves the **cavernous sinus**, and passes anteriorly through the **optic canal** with the **optic nerve** to enter the orbit. The ICA gives rise to the **anterior** and **middle cerebral arteries** and the **posterior communicating artery** of the circle of Willis.

The **pericallosal artery** is the main distal branch of the anterior cerebral artery which runs along the superior margin of the corpus callosum.

Exam tip:
- Remember that a proportion of the questions will be related to the structure marked with an arrow, but will not directly ask for its' name (e.g. Q3 above). Make sure to read each question!

Q1.17 Axial section of a T1-weighted sequence of an MRI brain

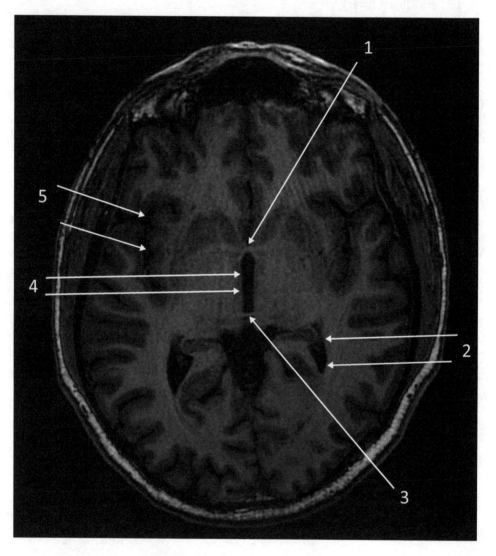

1. Name the arrowed structure.
2. Name the arrowed structure.
3. Name the arrowed structure.
4. Name the arrowed structure.
5. Name the arrowed structure.

Answers
1. Anterior commissure.
2. Temporal horn of the left lateral ventricle.
3. Posterior commissure.
4. Third ventricle.
5. Right Sylvian fissure (lateral sulcus).

Comments:
There are several white matter tracts which connect the right and left cerebral hemispheres, which are known as commissures. The largest of these is the **corpus callosum**. The **anterior commissure** is located anterior to the **third ventricle** on axial slices, and posterior to the **interhemispheric fissure**. On coronal sections it is inferior to the columns of the **fornix** and the **frontal horns of the lateral ventricles**. On midline sagittal slices it is anterosuperior to the third ventricle and is connected inferiorly to the **optic chiasm** via the **laminal terminalis**.

The **posterior commissure** is a small white matter tract which is posterior to the third ventricle on axial sequences. On midline sagittal sequences, it is a small ovoid structure superior to the **tectal plate** of the midbrain, and anterior to the **pineal gland** (and connected to both via the inferior pineal lamina).

Exam tip:
- The anterior and posterior commissures may not be visible on the same axial slice as in this image, as there will be variation in the way the images are acquired. They can be distinguished by their relation to the third ventricle.

Q1.18 Coronal section of a T1-weighted sequence of an MRI brain with gadolinium contrast

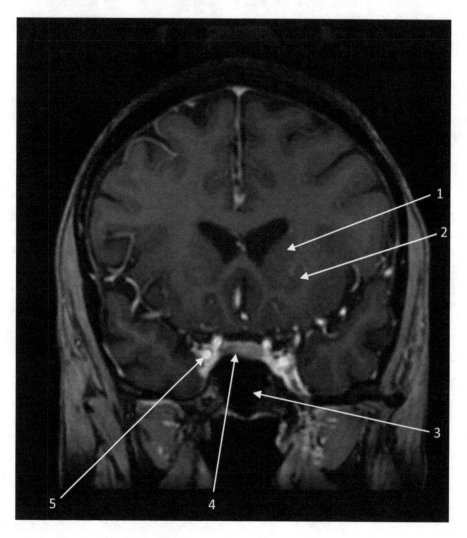

1. Name the arrowed structure.
2. Name the arrowed structure.
3. Name the arrowed structure.
4. Name the arrowed structure.
5. Name the arrowed structure.

Answers
1. Left caudate nucleus.
2. Left putamen.
3. Sphenoid sinus.
4. Pituitary gland.
5. Right internal carotid artery (cavernous segment).

Comments:
The **cavernous sinuses** are paired structures lateral to the **pituitary fossa**, and directly behind the apex of the orbits. They receive blood from a number of veins of the skull base.

The **internal carotid artery** passes through the cavernous sinus. The internal carotid artery is visualised twice on each side on this image, because it bends back on itself as it passes through the sinus.

Several cranial nerves also pass through the cavernous sinus. Cranial nerves III (oculomotor), IV (trochlear), Va (ophthalmic division of the trigeminal) and Vb (maxillary division of the trigeminal) are found on the lateral wall, and cranial nerve VI (abducens) passes through the middle of the sinus.

Exam tip:
- Be as specific as you can with your answers. Although arrow 4 is within the cavernous sinus, it is pointing directly at the internal carotid artery, so this is the answer that will give you full marks.

Q1.19 Heavily T2-weighted axial MRI slice of the posterior fossa

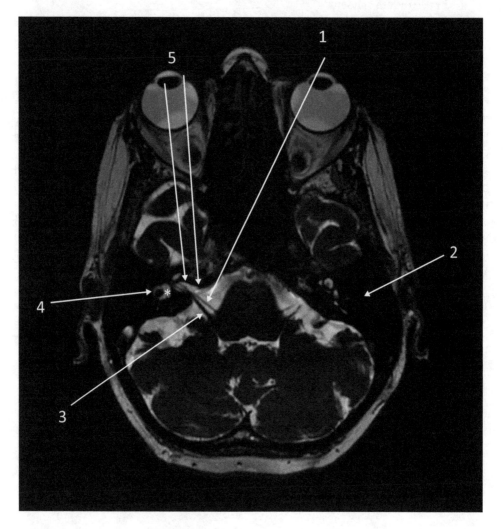

1. Name the arrowed structure.
2. Name the arrowed structure.
3. Name the arrowed structure.
4. Name the arrowed structure.
5. Name the arrowed structure.

Answers

1. Right facial nerve.
2. Left petrous portion of the temporal bone.
3. Right vestibulocochlear nerve.
4. Right lateral (horizontal) semicircular canal.
5. Right internal auditory canal (internal auditory/acoustic meatus).

Comments:

The image for this question is at the level of the inferior pons. The **facial nerve** arises slightly anterior to the **vestibulocochlear nerve**. Both pass into the **internal auditory canal** within the **petrous portion of the temporal bone**. The vestibulocochlear nerve splits into three. The configuration of these three nerves and the facial nerve within the internal auditory canal is shown below:

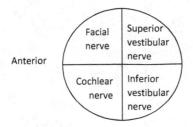

There are three **semicircular canals** which converge on the vestibule (marked with a yellow asterisk in this question). The lateral semi-circular canal is the smallest of the three and runs approximately in the axial plane. The anterior and posterior semicircular canal are at right angles to this.

Exam tip:

- Distinguishing the facial from the vestibulocochlear nerves is an exam classic—remember that the facial nerve arises more anteriorly from the pons.

Q1.20 Coronal section of a fluid-suppressed T2 MRI sequence of the brain

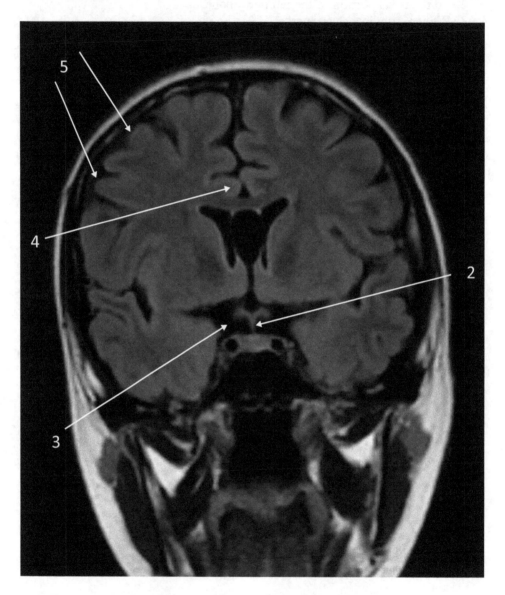

1. Name the anatomical variant.
2. Name the arrowed structure.
3. Name the arrowed structure.
4. Name the arrowed structure.
5. Name the arrowed structure.

Answers
1. Cavum septum pellucidum.
2. Pituitary infundibulum (stalk).
3. Suprasellar cistern.
4. Right cingulate gyrus.
5. Right middle frontal gyrus.

Comments:
The lateral surface of the **frontal lobe**, which lies under the skull vault, is divided into three gyri running in parallel. These are the **superior frontal, middle fontal** and **inferior frontal gyri**. The superior and middle gyri are separated by the **superior frontal sulcus**, and the middle and inferior gyri are separated by the **inferior frontal sulcus**. Posterior to these gyri is the **pre-central sulcus**, beyond which lies the **pre-central gyrus**, which is the most posterior part of the frontal lobe and contains the motor strip.

The **cingulate gyri** lie above the **corpus callosum**. Superior to the cingulate gyri are the **cingulate sulci**.

The **cavum septum pellucidum** and **cavum vergae** are anatomical variants in which there is a CSF space between the two leaflets of the septum pellucidum. They are defined by the anterior pillars of the fornices—the cavum septum pellucidum lies anterior to them and the cavum vergae posterior.

Exam tip:
- In the exam, you are most likely to see this anatomical variant in the axial plane. Although technically defined by the fornices, you won't be able to use this landmark in either axial or coronal planes. However, an anterior location indicates that it is a cavum septum pellucidum. The image below is from the same patient as the question, and demonstrates that this the CSF space actually spans the anterior and posterior portions—you would describe this as **cavum septum pellucidum et vergae**.

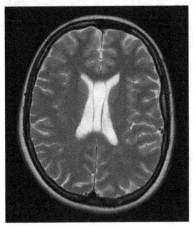

Q1.21 Digital subtraction angiogram of the posterior circulation of the brain

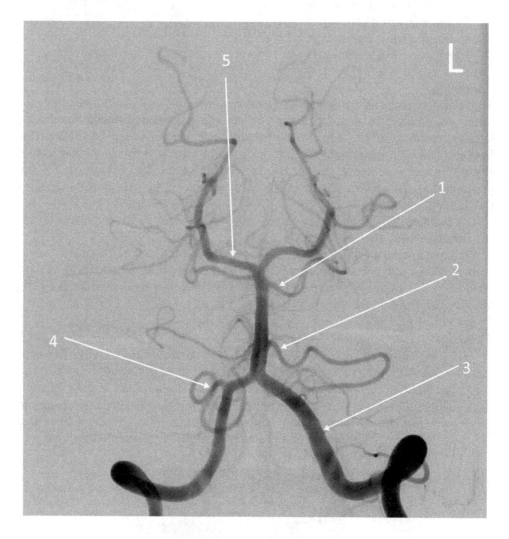

1. Name the arrowed structure.
2. Name the arrowed structure.
3. Name the arrowed structure.
4. Name the arrowed structure.
5. Name the arrowed structure.

Answers
1. Left superior cerebellar artery.
2. Left anterior inferior cerebellar artery.
3. Left vertebral artery.
4. Right posterior inferior cerebellar artery.
5. Right posterior cerebral artery.

Comments:

The **vertebral arteries** arise from the **subclavian arteries** and join to form the **basilar artery** anterior to the lower pons. The basilar artery usually divides to form the **posterior cerebral arteries**. In some patients, one or both posterior cerebral arteries may be supplied by dominant posterior communicating arteries (an anatomical variant—**foetal origin of the posterior cerebral artery**) and the first section of the posterior cerebral artery may be hypoplastic or absent.

Three main sets of paired arteries arise from the vertebrobasilar system to supply the cerebellum—the **posterior inferior cerebellar arteries**, the **anterior inferior cerebellar arteries** and the **superior cerebellar arteries**. There are anatomical variants in these vessels, but the most common configuration is described below. The posterior inferior cerebellar arteries are most inferior, and usually arise from the vertebral arteries. The anterior inferior cerebellar arteries arise next, from the basilar artery. The superior cerebellar arteries also arise from the basilar artery, close to its terminal bifurcation into the posterior cerebral arteries.

Exam tip:
- Remember that sometimes the left and right sides of the image are labelled. Keep an eye out, particularly when laterality isn't clear from the image alone.

Q1.22 Coronal section of a T1-weighted MRI sequence of the brain

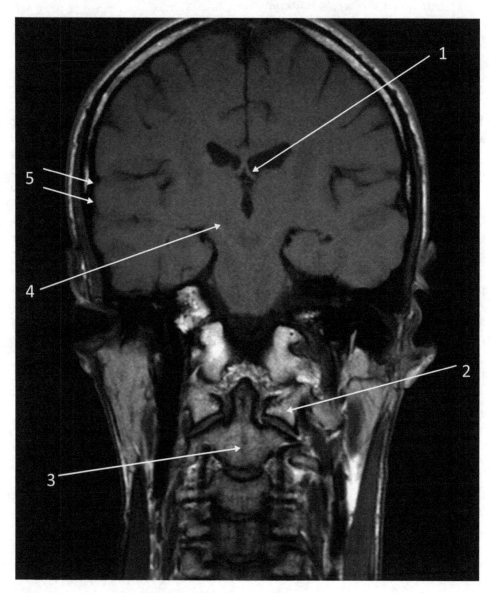

1. Name the arrowed structure.
2. Name the arrowed structure.
3. Name the arrowed structure.
4. Name the arrowed structure.
5. Name the arrowed structure.

Answers
1. Fornix.
2. Left lateral mass of C1.
3. Body of C2.
4. Right cerebral peduncle.
5. Right superior temporal gyrus.

Comments:
The **temporal lobes** are separated from the **frontal lobes** by the **lateral (Sylvian) fissures**. There are three main gyri running anteroposteriorly on the lateral surface of each temporal lobe. These are the **superior temporal gyrus**, the **middle temporal gyrus** and the **inferior temporal gyrus**. These are separated by the **superior** and **inferior temporal sulci**.

The **cerebral peduncles** form the anterior part of the **midbrain** and are the main communication between the brainstem and the thalamus and cortex.

Exam tip:
- Be specific if the arrows point to a specific part of a structure. Answering question 5 with 'temporal lobe' would likely get half marks.

Q1.23 T1 weighted axial MRI image of the brain with intravascular gadolinium contrast

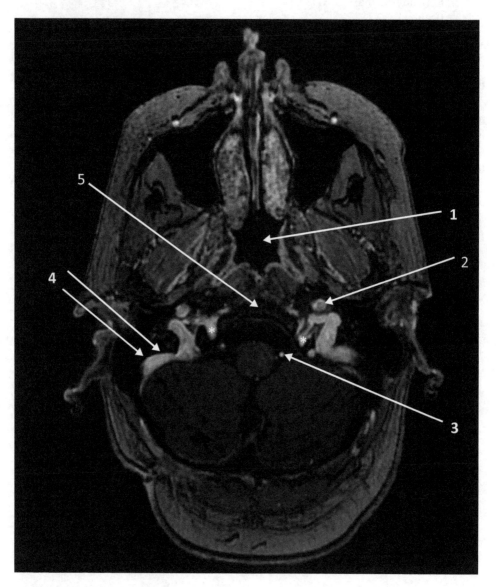

1. Name the arrowed structure.
2. Name the arrowed structure.
3. Name the arrowed structure.
4. Name the arrowed structure.
5. Name the arrowed structure.

Answers
1. Nasopharynx.
2. Left internal carotid artery.
3. Left vertebral artery.
4. Right sigmoid sinus.
5. Clivus.

Comments:
The **superior sagittal sinus** meets the **right** and **left transverse sinuses** at the **confluence of the sinuses (torcular herophili)**. The transverse sinuses become the **sigmoid sinuses**, which leave the skull base through the **jugular foramen** to form the **internal jugular veins**. The sigmoid sinus becomes the internal jugular vein when it is joined by the inferior petrosal sinus in the jugular foramen (the inferior petrosal sinuses are marked with yellow asterisks on the image—but note that they are beyond the necessary level of detail and very unlikely to come up in the exam).

The clivus is a backward-sloping structure in the skull base which is formed by part of the sphenoid bone superiorly and part of the **occipital bone** inferiorly. Posterior to the clivus sits the basilar artery and the pons. The inferior tip of the clivus is an anatomical landmark known as the **basion**, which is the anterior margin of the **foramen magnum**.

Exam tip:
- You may be more used to identifying the clivus on sagittal images. You can work out that it is the clivus on this axial image because it is a midline bony structure anterior to the medulla.

Q1.24 T1 weighted axial MRI image of the brain with intravascular gadolinium contrast

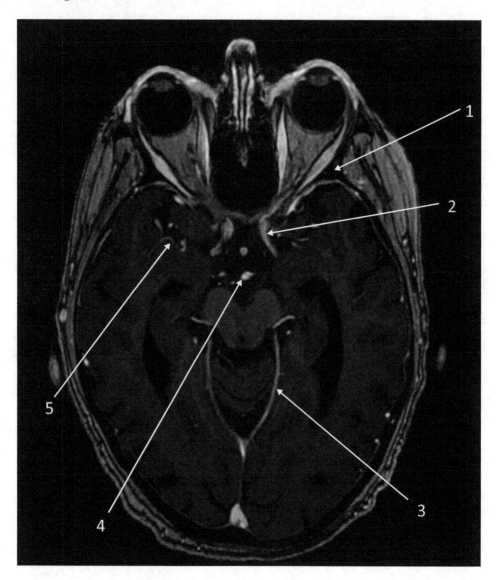

1. Name the arrowed structure.
2. Name the arrowed structure.
3. Name the arrowed structure.
4. Name the arrowed structure.
5. Name the arrowed structure.

Answers
1. Greater wing of the left sphenoid bone.
2. Left internal carotid artery.
3. Left tentorium cerebelli.
4. Basilar artery.
5. Right middle cerebral artery.

Comments:

The Circle of Willis is supplied by both the **internal carotid arteries** and the **vertebral arteries**. The internal carotid arteries give rise to the **anterior cerebral arteries**, the **middle cerebral arteries** and the **posterior communicating arteries**. The anterior cerebral arteries are connected by the **anterior communicating artery** after their first segments. The middle cerebral arteries travel laterally through the **lateral sulci (Sylvian fissures)**.

The **basilar artery** divides into two **posterior cerebral arteries**. These connect with the anterior circulation via the posterior communicating arteries.

Exam tip:
- This is another question in which the information given in the title may be helpful. The images are acquired with intravascular contrast, and bright structures are likely to represent vessels. However, you need to be slightly careful—the tentorium also appears bright on this sequence, and the bright structure in the middle of the suprasellar cistern is the pituitary infundibulum.

Q1.25 Midline sagittal section from an MRI of the brain

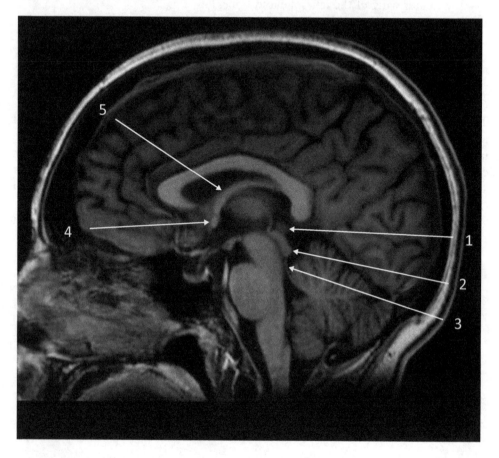

1. Name the arrowed structure.
2. Name the arrowed structure.
3. Name the arrowed structure.
4. Name the arrowed structure.
5. Name the arrowed structure.

Answers
1. Pineal gland.
2. Inferior colliculus.
3. Superior medullary velum.
4. Anterior commissure.
5. Fornix.

Comments:
The **superior cerebellar peduncles** are white matter tracts connecting the **midbrain** and **cerebellum**. The **superior medullary velum** is a thin tissue layer that spans the space between the two superior cerebellar peduncles, and forms part of the roof of the **fourth ventricle**.

The **fornix** is a white matter tract that arises posteriorly from the hippocampus. One crux of the fornix arises from each hippocampus, and curls superiorly and anteriorly until it joins with the contralateral crux. These form the **body of the fornix** which lies immediately below the septum pellucidum. As it continues anteriorly, the fornix splits again into two columns. Each **column** travels in an inferoposterior arc behind the **anterior commissure** and then lateral to the **third ventricle**. It then meets the ipsilateral **mamillary body**. The mamillary bodies lie at the base of the brain below the third ventricle and are seen on MRI anterior to the midbrain.

Exam tip:
- There are lots of structures that can be examined on a midline sagittal view, so it's worth taking time to familiarise yourself with it.

Q1.26 AP Skull Radiograph

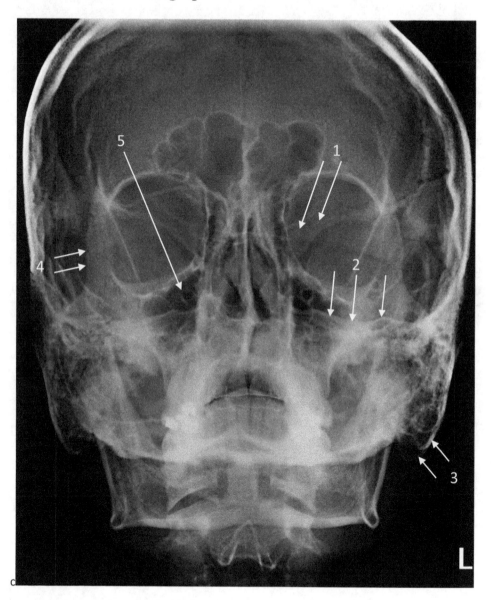

1. Name the arrowed structure.
2. Name the arrowed structure.
3. Name the arrowed structure.
4. Name the arrowed structure.
5. Name the arrowed structure.

Answers
1. Left lesser wing of the sphenoid bone.
2. Petrous ridge.
3. Left Mastoid process.
4. Frontal process of the right zygomatic bone.
5. Right foramen rotundum.

Comments:
A number of facial bones can be identified on an AP radiograph, including the **mandible**, the **maxilla**, the **zygomatic bones**, the **nasal bones** and parts of the **sphenoid bone**. In addition, structures in the upper cervical spine, the **frontal bone** and parts of the **temporal bones** can be distinguished.

The **zygomatic bone** articulates with the maxillary bone medially. It has two main processes; the **frontal process**, which is angled superiorly and articulates with the frontal bone at the **frontozygomatic suture**, and the temporal process which is angled posteriorly and articulates with the **zygomatic process of the temporal bone** to form the **zygomatic arch**.

The **sphenoid bone** has a complex structure. The **body of the sphenoid** attaches to a **greater wing** (inferiorly) and a **lesser wing** (superiorly) on each side; between each pair of wings is a **superior orbital fissure**. Several important foramina pass through the greater wings of the sphenoid, including the **foramen rotundum** and **foramen ovale**. A pterygoid process arises from the junction of the body of the sphenoid and the greater wing on each side, projecting downwards and splitting into **lateral** and **medial pterygoid plates**. The body of the sphenoid and pterygoid process are not readily visible on AP radiographs.

Exam tip:
- It is worth spending some time familiarising yourself with different facial views, because there are many structures that can be examined, and because different projections can make familiar structures more difficult to identify.

Q1.27 Skull Radiograph in an occipitomental projection

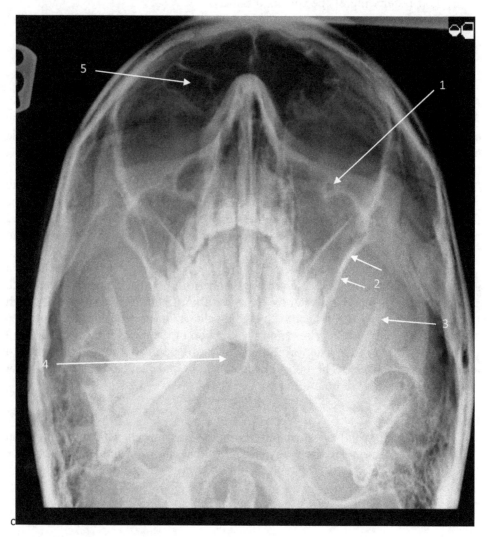

1. Name one structure that passes through this foramen.
2. Name the arrowed structure.
3. Name the arrowed structure.
4. Name the arrowed structure.
5. Name the arrowed structure.

Answers
1. Left infraorbital nerve, artery or vein (left infraorbital foramen).
2. Lateral wall of the left maxillary sinus.
3. Left coronoid process of the mandible.
4. Right Sphenoid sinus.
5. Right frontal sinus.

Comments:
There are four pairs of paranasal sinuses—the **maxillary, frontal, sphenoid** and **ethmoid** sinuses. The maxillary sinuses are the largest of the paranasal sinuses, and are pyramidal in shape with **lateral, medial** and **anterior walls.** The **orbital floor** is the roof of the maxillary sinus.

The ethmoid air cells (sinuses) are situated between the upper nasal cavity medially and the orbit laterally. They are separated from the orbits by a strip of bone called the **lamina papyracea.** Posterior to the upper nasal cavity and ethmoid air cells are the sphenoid sinuses within the sphenoid bone.

The mandible comprises a **body** medially, which meets a more vertically-orientated **ramus** on each side at the **angles** of the mandible. Arising from each ramus is a **condylar process** which articulates with the mandibular fossa at the temporomandibular joint, and a **coronoid process of the mandible**, which serves as an insertion for the **temporalis** and **masseter muscles**.

Exam tip:
- The sphenoid sinus question is particularly tricky, because it appears to be more inferior than expected due to the projection. If you couldn't initially recognise it on this radiograph, you could deduce the answer from observation that it is a paired air-filled structure near to the midline.

Q1.28 Transverse section from an ultrasound of the floor of the mouth

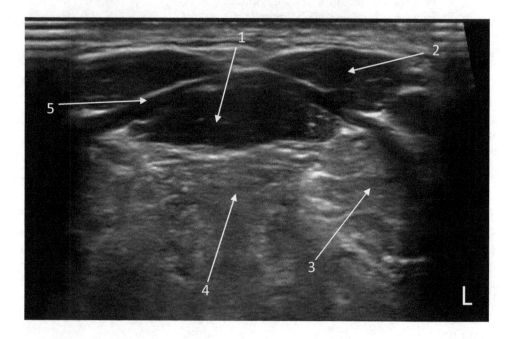

1. Name the arrowed structure.
2. Name the arrowed structure.
3. Name the arrowed structure.
4. Name the arrowed structure.
5. Name the arrowed structure.

Answers
1. Geniohyoid muscle.
2. Anterior belly of the left digastric muscle.
3. Left sublingual gland.
4. Genioglossus muscle.
5. Right mylohyoid muscle.

Comments:
Each **digastric muscle** has an anterior and posterior belly—the **anterior belly** inserts to the **mandible**.

The **mylohyoid muscle** is a triangular structure that lies beneath the anterior belly of the digastric, between the mandible and **hyoid bone**. It is a useful landmark on MRI because it separates the submandibular and sublingual spaces.

The **geniohyoid muscles** are paired midline structures (although they are inseparable in the image above) which also run between the mandible and hyoid. Immediately above them is the **genioglossus muscle**, which is a large intrinsic tongue muscle. The prefix 'genio' refers to the chin.

The **sublingual glands** are one of three paired salivary glands, the others being the **submandibular glands** and the **parotid glands**.

Exam tip:
- The title is particularly helpful when given ultrasound images as it helps you to orientate yourself to the correct part of the body.

Q1.29 Transverse section from an ultrasound examination of the neck

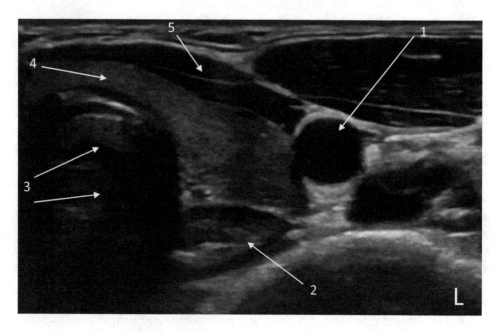

1. Name the arrowed structure.
2. Name the arrowed structure.
3. Name the arrowed structure.
4. Name the arrowed structure.
5. Name the arrowed structure.

Answers
1. Left common carotid artery.
2. Oesophagus.
3. Trachea.
4. Isthmus of the thyroid gland.
5. Left strap muscles.

Comments:
The **thyroid gland** is positioned in the visceral compartment of the neck, enveloped by the pretracheal fascia. It lies behind the **strap muscles**, and inferior to the **thyroid cartilage**. The two **lobes** of the thyroid gland are situated on either side of the **trachea** and connected by an **isthmus** anteriorly.

The **sternocleidomastoid** muscle anatomically divides the neck into anterior and posterior triangles. It originates from the middle third of the **clavicle** and **manubrium of the sternum**, and inserts onto the **mastoid process of the temporal bone**.

The left and right **common carotid arteries** are branches of the aortic arch and brachiocephalic trunk, respectively. Both vessels run from the sternoclavicular joint to the upper border of the thyroid cartilage and bifurcate at the level of C3/C4, into the **external** and **internal carotid arteries**. At the level of the thyroid, the **internal jugular vein** lies lateral to the common carotid artery, and higher in the neck it is positioned posteriorly.

Exam tip:
- Several views on neck ultrasound are more likely to be shown than others because they allow identification of structures on a single plane—particularly the floor of the mouth, thyroid, and Doppler images of the vascular structures.

Q1.30 Single image taken from a dacryocystogram study

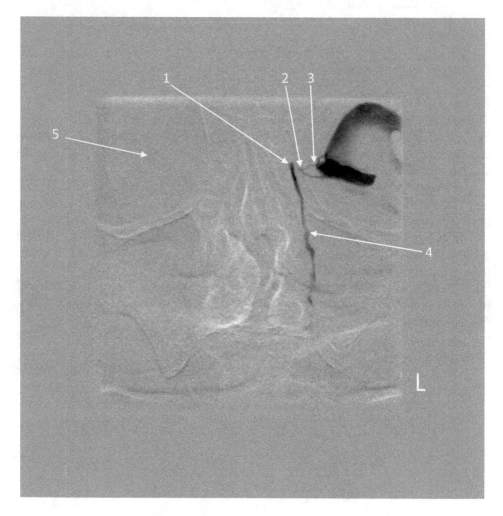

1. Name the arrowed structure.
2. Name the arrowed structure.
3. Name the arrowed structure.
4. Name the arrowed structure.
5. Name the arrowed structure.

Answers

1. Left lacrimal sac.
2. Left common canaliculus.
3. Left superior canaliculus.
4. Left nasolacrimal duct.
5. Right orbit.

Comments:

A dacryocystogram, a fluoroscopic examination of the nasolacrimal ducts, is conducted by inserting a catheter and injecting contrast. The **superior** and **inferior canaliculi** drain tears from the medial canthus of the eye, and join to form the **common canaliculus**. This drains into the **lacrimal sac**, which in turn drains into the **nasolacrimal duct** below the inferior orbital margin. The nasolacrimal duct opens into the **inferior nasal meatus**.

The coronal CT image of the skull below demonstrates the location of the right lacrimal sac (indicated by a yellow asterisk) and the start of the right nasolacrimal duct (indicated by a blue cross).

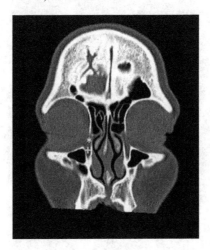

Exam tip:

- This is a difficult question because it is an examination that you're unlikely to have encountered in clinical practice. Take a moment to familiarise yourself with the key structures on parotid and submandibular sialograms as well.

Q1.31 Coronal T1-weighted sequence from a MRI of the head

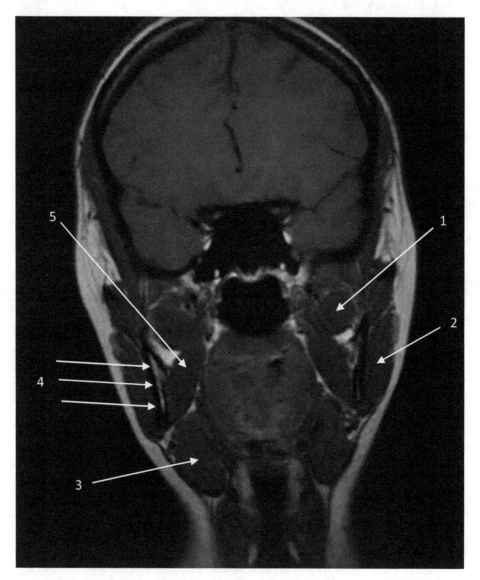

1. Name the arrowed structure.
2. Name the arrowed structure.
3. Name the arrowed structure.
4. Name the arrowed structure.
5. Name the arrowed structure.

Answers
1. Left lateral pterygoid muscle.
2. Left masseter muscle.
3. Right submandibular gland.
4. Right ramus of the mandible.
5. Right medial pterygoid muscle.

Comments:
The **masseter, temporalis, medial pterygoid** and **lateral pterygoid muscles** together comprise the muscles of mastication, involved in moving the jaw during chewing.

The master muscle attaches to the inner surface of the **ramus and coronoid process of the mandible**, and to the **zygomatic arch**.

The temporalis is a large flat muscle which originates at the **parietal bone** of the skull and dives forward beneath the zygomatic arch to insert into the coronoid process of the mandible.

The pterygoid muscles attach to the **lateral pterygoid plate**; the lateral pterygoid muscle attaches to the **condylar process** and the medial pterygoid muscle to the ramus of the mandible.

The lateral pterygoid plate and **medial pterygoid plate** are part of the **sphenoid bone** and are shown on the coronal CT section displayed below.

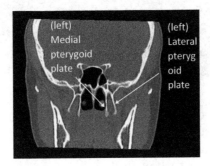

Exam tip:
- Remember to include descriptors such as 'muscle', 'tendon', 'artery' in your answer—if you don't you may lose marks.

Q1.32 Coronal T1-weight sequence from a MRI of the neck

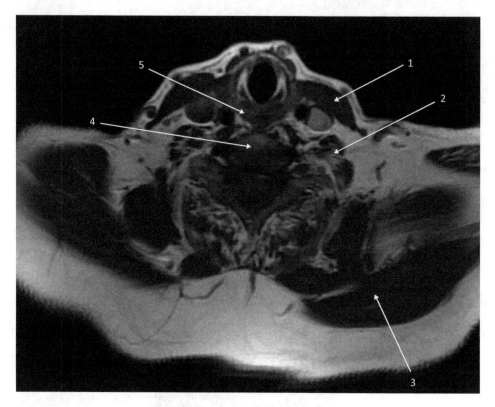

1. Name the arrowed structure.
2. Name the arrowed structure.
3. Name the arrowed structure.
4. Name the arrowed structure.
5. Name the arrowed structure.

Answers
1. Left sternocleidomastoid muscle.
2. Left scalene muscles.
3. Left trapezius muscles.
4. Cervical vertebral body.
5. Oesophagus.

Comments:
There are three **scalene muscles** in the lateral neck. The anterior and middle scalene muscles elevate the first rib, and the posterior scalene muscle elevates the second rib. The **subclavian vein** and phrenic nerve lie in front of the anterior scalene muscle. The nerves of the brachial plexus and the **subclavian artery** lie between the anterior and middle scalene muscles.

The **trapezius** is a large muscle which stretches from the **occipital bone** to the **spinous process of T12** in the midline and inserts onto the **clavicle** and **scapula**. It is superficial to other muscles in the posterior neck and upper back.

Exam tip:
- Many neck muscles are difficult to distinguish on individual cross-sectional images and are therefore unlikely to come up in the exam. Focus on learning important muscles (such as the sternocleidomastoid and trapezius) and identifiable groups of muscles (such as the strap and scalene muscles).

Q1.33 Axial CT of the base of skull

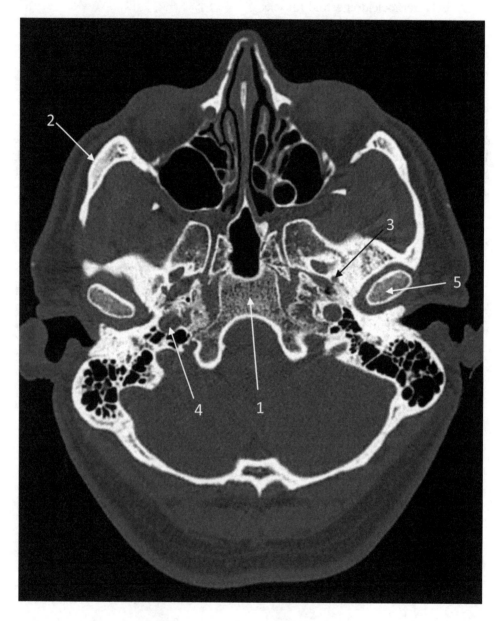

1. Name the arrowed structure.
2. Name the arrowed structure.
3. Name one structure that runs through this foramen.
4. Name one structure that runs through this foramen.
5. Name the arrowed structure.

Answers
1. Body of the sphenoid bone.
2. Right zygomatic arch.
3. Left middle meningeal artery, middle meningeal vein (left foramen spinosum).
4. Right internal carotid artery (carotid canal).
5. Left mandibular condyle.

Comments:
The cranial foramina are holes that allow the passage of structures (nerves, vessels etc.) through the base of skull. They are important anatomical sites where these structures are prone to injury and have surgical implications. A summary of key cranial foramina is given in the table below:

Cranial foramen	Traversing structure(s)
Foramen ovale	Mandibular division of the trigeminal nerve (V3), accessory meningeal artery
Foramen spinosum	Middle meningeal artery, middle meningeal vein
Foramen lacerum	Ascending pharyngeal artery, greater petrosal nerve
Carotid canal	Internal carotid artery
Jugular foramen	Pars nervosa: glossopharyngeal nerve (IX)Pars vascularis: vagus nerve (X), spinal accessory nerve (XI), Jugular bulb (first part of the internal jugular vein)
Foramen magnum	Spinal cord, vertebral arteries, anterior and posterior spinal arteries, spinal root of accessory nerve (XI)
Hypoglossal canal	Hypoglossal nerve (XII)
Stylomastoid foramen	Facial nerve (VII)

Exam tip:
- The cranial foramina can be difficult. Remember that some foramina, particularly the foramen rotundum, are orientated in the axial plane and are best identified on coronal sections.

IOP Publishing

Anatomy for the Royal College of Radiologists Fellowship
Illustrated questions and answers

Andrew G Murchison, Mitchell Chen, Thomas Frederick Barge, Shyamal Saujani, Christopher Sparks, Radoslaw Adam Rippel, Malcolm Sperrin and Ian Francis

Chapter 2

Thorax

Mitchell Chen

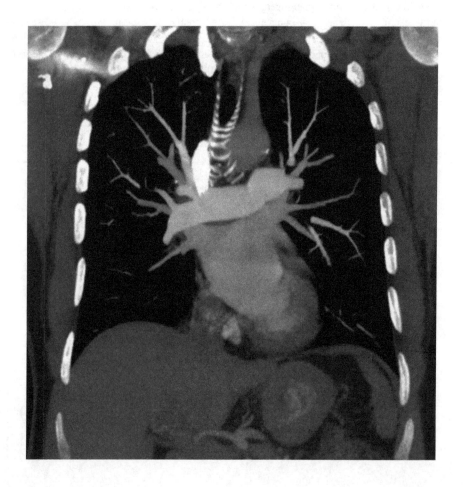

Q2.1 Frontal chest radiograph

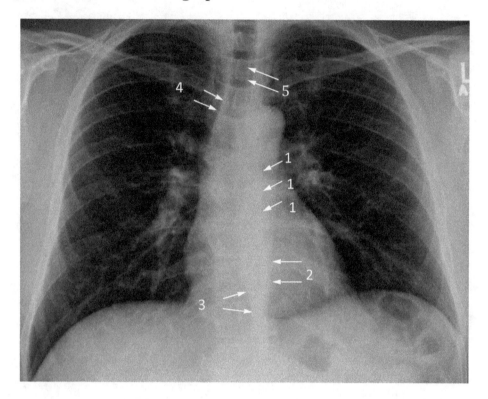

1. Name the arrowed structure.
2. Name the arrowed structure.
3. Name the arrowed structure.
4. Name the arrowed structure.
5. Name the arrowed structure.

Answers
1. Left paraspinal line.
2. Descending aorta (para-aortic line).
3. Azygo-oesophageal line.
4. Right paratracheal stripe.
5. Posterior junctional line.

Comments:

The **paraspinal lines** represent the interface between the lungs and posterior mediastinal soft tissues. The left paraspinal line is more commonly seen than the right and lies medial to the lateral border of the descending aorta.

The **Azygo-oesophageal line** marks the medial border of the azygo-oesophgeal recess (see question 3).

The **paratracheal stripes** represent the tracheal walls, adjacent pleural surfaces and any mediastinal fat between them. They are abnormal if their thickness is greater than 4 mm.

The **posterior junctional line** is formed from the apposition of the pleural surfaces of the upper lobes of the lungs and lies in between the oesophagus anteriorly and thoracic spine posteriorly. The anterior junctional line is the equivalent in the anterior mediastinum and lies more inferiorly on frontal chest radiographs.

Exam tip:
- Not all lines are seen on all plain radiographs. It's important to familiarise yourself with the position of different lines, especially ones that are in close proximity.

Q2.2 Frontal chest radiograph

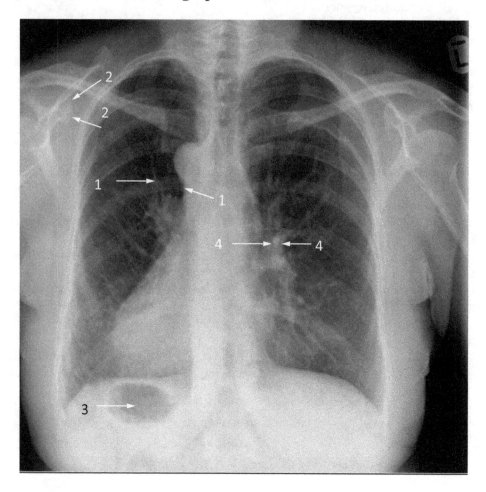

1. Name the arrowed anatomical area.
2. Name the arrowed structure.
3. Name the arrowed structure.
4. Name the arrowed structure.
5. Name the anatomical variant.

Answers
1. Aorto-pulmonary window.
2. Coracoid process of the right scapula.
3. Stomach gas bubble.
4. Lobar branch of the left pulmonary artery supplying the left upper pulmonary lobe.
5. Situs Inversus.

Comments:
Note the right-sided heart, stomach and the left-sided liver in this case. If only the heart is involved, the answer would be dextrocardia. Situs inversus is congenital condition where major visceral organs are found on the opposite side to their normal positions (a normal configuration of organs is known as situs solitus). Situs inversus associated with primary ciliary dyskinesia is termed Kartagner's syndrome, and is clinically characterised by a triad: situs inversus, chronic sinusitis, and bronchiectasis.

Q2.3 Axial cardiac coronary angiography

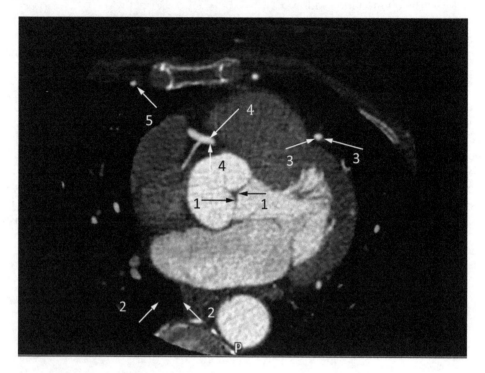

1. Name the arrowed structure.
2. Name the arrowed structure.
3. Name the arrowed structure.
4. Name the arrowed structure.
5. Name the arrowed structure.

Answers

1. Aortic valve.
2. Azygo-oesophageal recess.
3. Left anterior descending artery.
4. Right main coronary artery.
5. Right internal thoracic artery/right internal mammary artery.

Comments:

The left and **right coronary arteries** arise from the **left** and **right coronary sinuses**, respectively. They are the only two branches of the ascending aorta. The right coronary artery travels in the right atrioventricular groove to the inferior surface of the heart and become the posterior descending artery (PDA) in a right dominant circulation. The PDA is a branch of the left coronary artery in a left dominant circulation.

The **azygo-oesophageal recess** is a pre-vertebral space formed by the posteromedial segment of the right lower lobe, the **azygos vein** and oesophagus.

Exam tip:

- Coronary vasculature can be examined both in the form of coronary angiography or cardiac CT.

Q2.4 Coronal contrast-enhanced CT of the chest

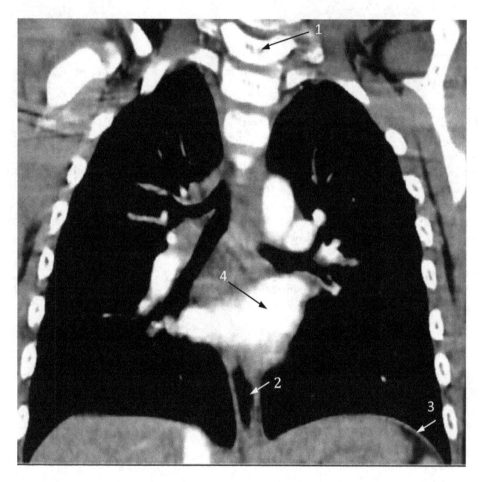

1. Name the arrowed structure.
2. Name the arrowed structure.
3. Name the arrowed structure.
4. Name the arrowed structure.
5. Name the anatomical variant.

Answers
1. Body of the 7th cervical vertebra.
2. Oesophagus.
3. Left hemidiaphragm.
4. Left atrium.
5. Tracheal bronchus (*bronchus suis*).

Comments:
The **left atrium** is the most posterior chamber of the heart and is directly inferior to the **carina** and main bronchi.

Tracheal bronchus, or ***bronchus suis*** ('pig bronchus') is an accessory bronchus originating directly from the supracarinal trachea. Cardiac bronchus is when the accessory bronchus originates from **bronchus intermedius**.

One way to identify the level of an arrowed vertebral body is to first identify the ribs, which are usually only found in the thoracic spine (with the exception of the anatomical variant **cervical ribs**). You can count up or down from the first vertebra that has an associated rib, which is T1, or the last, which is T12.

It can also be helpful to understand the differences in the shape of cervical, thoracic and lumbar vertebrae. The cervical vertebrae have bifid **spinous** and **transverse processes**, as well as two **transverse foramina** (through which the **vertebral arteries** pass). The lumbar vertebrae have more posteriorly-projecting spinous processes, compared to the inferiorly—projecting processes of the thoracic spine.

Exam tip:
- The exam can test structures at the periphery of the image, such as the T1 body in this case. Learn to count ribs and vertebra on different projections and planes.

Q2.5 T1-weighted coronal section from an MR arteriogram of the carotids with gadolinium contrast

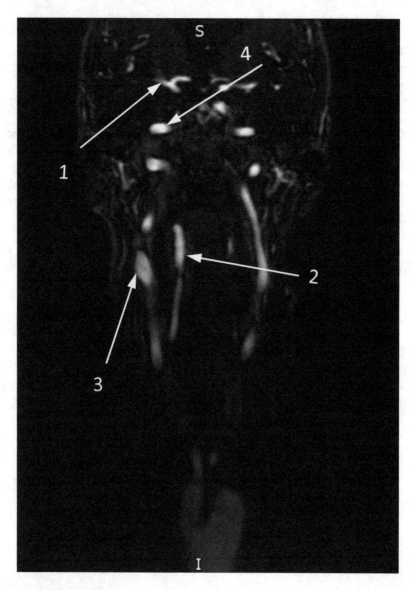

1. Name the arrowed structure.
2. Name the arrowed structure.
3. Name the arrowed structure.
4. Name the arrowed structure.
5. Name the anatomical variant.

Answers

1. M1 segment of the right middle cerebral artery.
2. Cervical segment of the right internal carotid artery.
3. Right external carotid artery.
4. Cavernous segment of the right internal carotid artery.
5. Bovine aortic arch.

Comments:

The hyperintense vascular structures demonstrated on this image are arterial, as suggested by the title of the question. The **common carotid artery** bifurcates into the **external** and **internal carotid artery** at the level of C4. The segments of the internal carotid artery are cervical, petrous, lacerum, cavernous, clinoid, opthalmic and communicating (terminal). Its **cavernous segment** is located in the **cavernous sinus**.

The **aortic arch** at the bottom of the image is of lower signal because the MRI is timed to maximise enhancement of the carotid arterial branches. A **bovine arch** is the most common variant of the aortic arch, where the **brachiocephalic artery** shares a common origin with the **left common carotid artery**. In this example, this can be appreciated by observing that both common carotids arise from the same branch coming from the aortic arch.

Exam tip:

- If unsure about a normal variant question, it's best to sit back and think about the common variants that can be present in the body area shown in the image.

Q2.6 Axial unenhanced CT of the chest

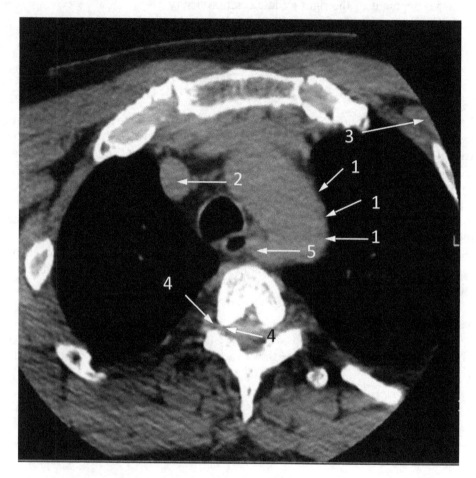

1. Name the arrowed structure.
2. Name the arrowed structure.
3. Name the arrowed structure.
4. Name the arrowed structure.
5. Name the anatomical variant.

Answers
1. Aortic arch.
2. Right brachiocephalic vein.
3. Left pectoralis minor muscle.
4. Right thoracic intervertebral/neural foramen.
5. Aberrant right subclavian artery.

Comments:
Branches of the aortic arch are the **brachiocephalic artery**, the **left common carotid artery** and the **left subclavian artery**.

The **right subclavian artery** usually arises from the brachiocephalic artery. An **aberrant right subclavian artery** is a vascular anomaly in which the right subclavian branches directly from the aortic arch. In most cases, it crosses from left to right posterior to the **oesophagus**.

Note that the answer for question 2 is the **right brachiocephalic vein** rather than the **superior vena cava** because you can also see the **left brachiocephalic vein** separately, passing anterior to the aorta. Therefore, the left and right brachiocephalic veins have not joined yet, so this cannot be the SVC.

Exam tip:
- It is worthwhile familiarising yourself with common anatomical variant configurations of the thoracic aorta.

Q2.7 Lateral plain radiograph of the chest

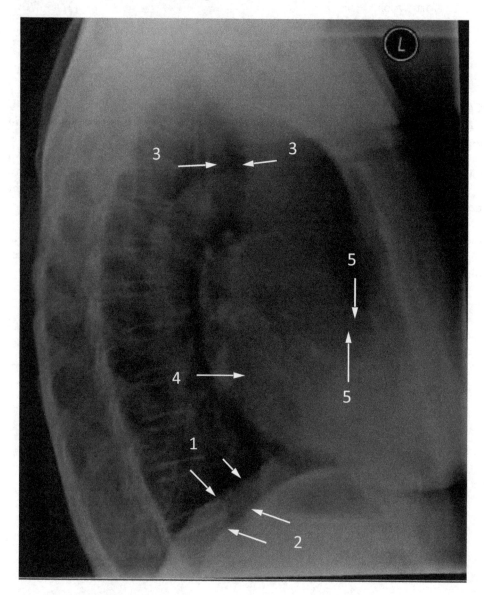

1. Name the arrowed structure.
2. Name the arrowed structure.
3. Name the arrowed structure.
4. Name the arrowed structure.
5. Name the arrowed structure.

Answers
   1. Right hemidiaphragm.
   2. Left hemidiaphragm.
   3. Trachea.
   4. Left atrium.
   5. Right ventricle.

Comments:
The most reliable way of distinguishing between the **right and left hemidiaphragms** on a lateral x-ray is that the right extends to the anterior chest wall whereas the left terminates at the posterior heart border.

The order of the heart chambers from anterior to posterior is **right ventricle, left ventricle, right atrium** and **left atrium**.

Exam tip:
   • A solid grasp of the 3D anatomy of the heart chambers and surrounding structures in the mediastinum is vital to interpreting chest x-rays.

Q2.8 Axial CT cardiac coronary angiogram

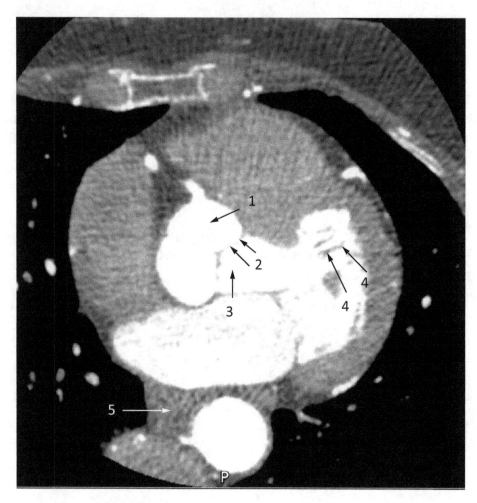

1. Name the arrowed structure.
2. Name the arrowed structure.
3. Name the arrowed structure.
4. Name the arrowed structure.
5. Name the arrowed structure.

Answers
1. Anterior aortic sinus.
2. Right coronary cusp.
3. Left ventricular outflow tract.
4. Papillary muscles in the left ventricle.
5. Oesophagus.

Comments:
The **anterior aortic sinus** can be identified because it is the origin of the **right coronary artery,** which is displayed on this image.

The normal aortic valve consists of three valve cusps (or leaflets): the right, left and posterior cusps. These are followed distally by three aortic sinuses: the **anterior, left** and **posterior aortic sinuses**. The anterior aortic sinus gives rise to the right coronary artery and the left aortic sinus give rise to the **left coronary artery**. The posterior aortic sinus is also called the non-coronary sinus because it does not give rise to any cardiac vessels.

Posterior to the **oesophagus** is the **azygous vein**. The oesophagus can be identified by its position and gas-containing lumen (partially collapsed in this case).

Q2.9 Coronal T2 sequence from a magnetic resonance cholangiopancreatography (MRCP) examination

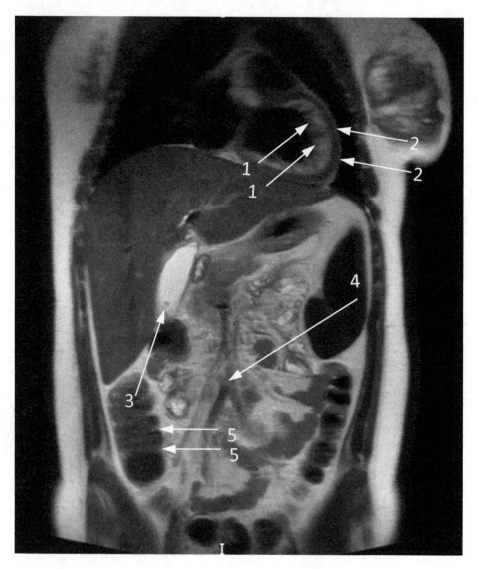

1. Name the arrowed structure.
2. Name the arrowed structure.
3. Name the arrowed structure.
4. Name the arrowed structure.
5. Name the arrowed structure.

Answers
1. Myocardium of the left ventricle.
2. Pericardium of the left ventricle.
3. Fundus of the gallbladder.
4. Bifurcation of the abdominal aorta.
5. Caecal haustra.

Comments:
The layers of the heart are the **pericardium, myocardium** and **endocardium**. In the absence of pathology, only the first two layers can be distinctly visualised on imaging. Assessing the appearance and perfusion of the myocardium is a key purpose of cardiac MR.

The **gallbladder** is hyperintense on this MRCP study, giving it a high sensitivity for biliary pathology.

Q2.10 Lateral plain radiograph of the sternum

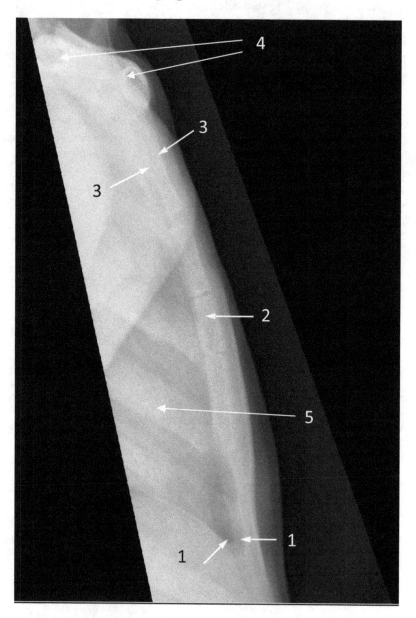

1. Name the arrowed structure.
2. Name the arrowed structure.
3. Name the arrowed structure.
4. Name the arrowed structure.
5. Name the arrowed structure.

Answers
1. Sterno-phrenic angle.
2. Ossification centre of the sternum.
3. Manubrium of the sternum.
4. Clavicle.
5. Anterior rib.

Comments:

The sterno-phrenic angle is formed by the sternum and the anterior diaphragm.

The sternum is composed of the **manubrium**, the **sternal body** and the **xiphisternum** (also known as the xiphoid process). However, in the paediatric population, the sternal body is made up of small rounded **ossification centres**. These are usually numerous, and fuse by early adulthood to form the body of the sternum.

Anteriorly the ribs articulate with **costal cartilage** at the costochondral joints. On a lateral study such as this, only the anterior aspects of the ribs are demonstrated.

Exam tip:
- You will only be tested on normal structures in FRCR Part I. If a pointed structure appears pathological, think about developmental phases or normal variants.

Q2.11 Axial contrast-enhanced CT of the chest

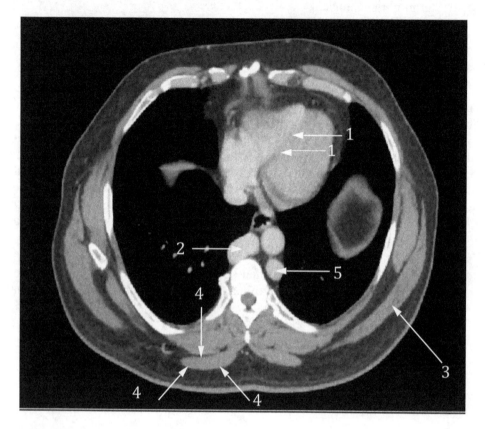

1. Name the arrowed structure.
2. Name the arrowed anatomical variant.
3. Name the arrowed structure.
4. Name the arrowed structure.
5. Name the arrowed anatomical variant.

Answers
1. Interventricular septum.
2. Azygos continuation of inferior vena cava.
3. Left latissimus dorsi muscle.
4. Right trapezius muscle.
5. Hemiazygos continuation of inferior vena cava.

Comments:
The **azygos** and **hemiazygos** veins are enlarged. The suprarenal **inferior vena cava** is absent and drainage is instead via the azygos and hemiazygos veins. Note that the hepatic veins still drain into the right atrium via the IVC.

Exam tip:
- It's important to remember that whilst no pathology is tested in this exam, it's still useful to have a working knowledge of common features and the normal size of structures to help identify normal variants.

Q2.12 Frontal chest radiograph

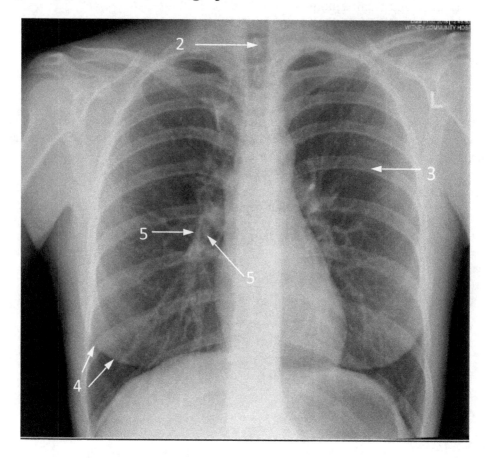

1. Name the anatomical variant.
2. Name the arrowed structure.
3. Name the arrowed structure.
4. Name the arrowed structure.
5. Name the arrowed structure.

Answers
1. Azygos fissure.
2. Spinous process of the T1.
3. Posterior left 6th rib.
4. Right breast shadow.
5. Right descending pulmonary artery.

Comments:
The **azygos fissure** is the most common accessory fissure of the lung.

Exam tips:
- There is no need to specify the side for unilateral structures, such as the horizontal or azygos fissures in this question.
- Where possible, specify whether the arrow is pointing to the posterior, lateral or anterior part of the rib.

Q2.13 Sagittal section from a CT of the chest

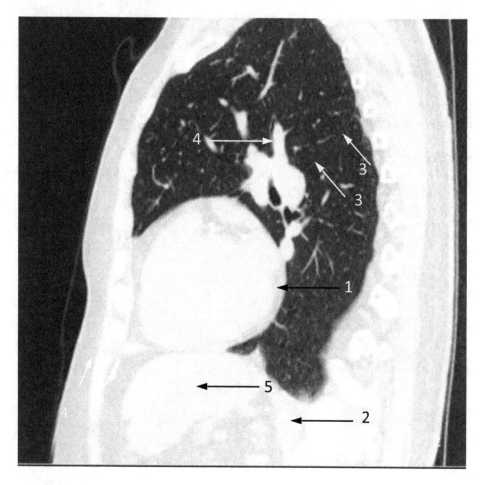

1. Name the arrowed structure.
2. Name the arrowed structure.
3. Name the arrowed structure.
4. Name the arrowed structure.
5. Name the arrowed structure.

Answers
1. Myocardium.
2. Spleen.
3. Left oblique fissure.
4. Left upper lobar pulmonary artery.
5. Left lobe of the liver.

Comments:
The **pulmonary trunk** bifurcates in the transthoracic plane of Ludwig, which runs from the manubrio-sternal joint to the inferior aspect of T4. The **left pulmonary artery** is shorter than the right and arches posteriorly over the **left main bronchus** and divides into upper (ascending) and lower (descending) lobar arteries at the left hilum. The left upper lobar pulmonary artery supplies the left upper lobe, and the left lower lobar pulmonary artery supplies the lingula of the left upper lobe and the left lower lobe.

We know that the arrowed structure in question 4 is a branch of the left pulmonary artery because it is posterior to the left superior pulmonary vein. The pulmonary veins are anterior and slightly inferior to the pulmonary arteries.

Exam tip:
- A working knowledge of broncho-pulmonary segments and their associated bronchi and vasculature is important for the exam, as it is a frequently-tested area.

Q2.14 Sagittal contrast-enhanced CT of the chest

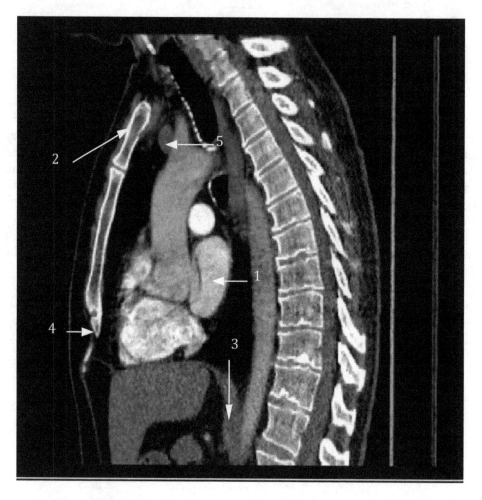

1. Name the arrowed structure.
2. Name the arrowed structure.
3. Name the arrowed structure.
4. Name the arrowed structure.
5. Name the arrowed structure.

Answers
1. Left atrium.
2. Manubrium.
3. Crus of the Left hemi-diaphragm.
4. Xiphisternum.
5. Left brachiocephalic vein.

Comments:
Right ventricle is the most anterior part of the heart and the **left atrium** the most posterior. The **crus of the diaphragm** is the muscle part that stretches in the abdomen and can be seen on cross-sectional abdominal imaging.

The **brachiocephalic veins** join with each other to form the superior vena cava, which then drain into the right atrium of the heart, receiving also from the inferior vena cava via the azygous vein.

Q2.15 Medial lateral oblique view mammogram

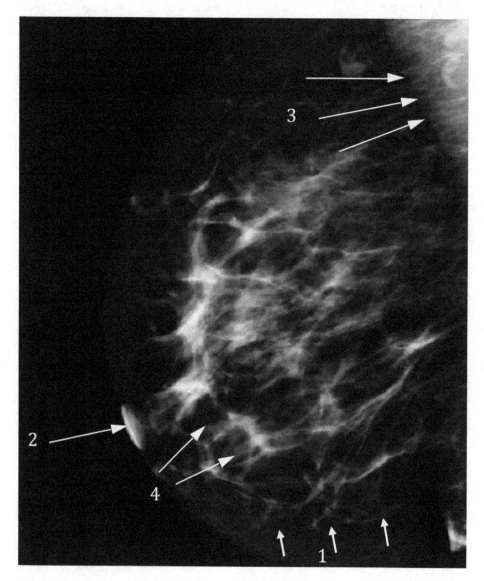

1. Name the arrowed structure.
2. Name the arrowed structure.
3. Name the arrowed structure.
4. Name the arrowed structure.

Answers
1. Cooper ligament.
2. Nipple-areolar complex.
3. Pectoralis major muscle.
4. Breast lobules.

Comments:
The **Cooper ligaments** are fibrous bands that connect the inner side of the breast skin to the **pectoral muscles**. The crests of Duret attach superficial **breast lobules** by their summits to the superficial breast fascia.

Exam tip:
- Remember to familiarise yourself with anatomy on modalities such as mammography and fluoroscopy—don't only focus on CT and MRI in your revision.

Q2.16 Axial computed tomography chest with contrast

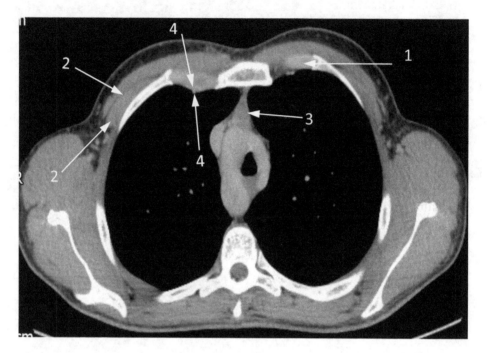

1. Name the arrowed structure.
2. Name the arrowed structure.
3. Name the arrowed structure.
4. Name the arrowed structure.
5. Name the anatomical variant.

Answers
1. Costal cartilage of a left rib.
2. Right pectoralis minor muscle.
3. Thymus.
4. Right internal thoracic artery (internal mammary artery).
5. Double aortic arch.

Comments:

The **costal cartilage** connects the true ribs with the sternum via the sternocostal joints. The false ribs are floating ribs that do not connect to the sternum.

The **thymus** is physiologically large in children and adolescents and then gradually involutes with age, with fatty replacement of the cellular component. It is triangular-shaped in adults with small blood vessel seen traversing it.

The **internal thoracic artery (internal mammary artery)** arises from the first part of the **subclavian artery** and is a key landmark to avoid when it comes to trans-thoracic interventions such as anterior-approach image-guided lung biopsies. It bifurcates at the level of 6th/7th anterior rib into the muculophrenic artery and superior epigastric artery. The name mammary comes from the fact that it gives off perforating branches to supply the breasts. Note that the **inferior epigastric artery** (a key landmark in inguinal anatomy) is a branch of the **external iliac artery** which anastomoses with the superior epigastric artery to supply the anterior abdominal wall.

The **double aortic arch** is the most common symptomatic aortic arch variant. It is due to persistent embryonic fourth aortic arches. It is right-dominant in 75%–80%, left dominant in 25% and co-dominant in 5% of cases. In a contrast swallow study, indentations can be seen anterior to the trachea and posterior to the oesophagus.

Exam tip:
- Commonly-tested normal variants that you should be familiar with include: right-sided aorta, double aortic arch, aberrant subclavian artery and aberrant left pulmonary artery.

Q2.17 Axial section from a computed tomography pulmonary angiogram (CTPA)

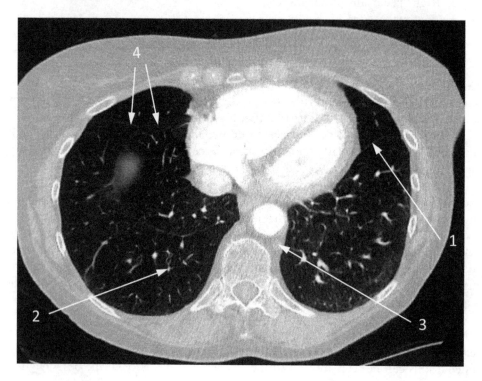

1. Name the arrowed structure.
2. Name the arrowed structure.
3. Name the arrowed structure.
4. Name the arrowed structure.
5. Name the anatomical variant.

Answers
1. Left oblique fissure (major fissure).
2. Pulmonary vessel in the posterior basal segment of the right lower lobe.
3. Hemi-azygous vein.
4. Horizontal fissure (minor fissure).
5. Inferior accessory fissure of the right lung (Twining's line).

Comments:
In normal anatomy, each lung has ten bronchopulmonary segments. Lung fissures are double-folded layers of visceral pleura that are usually only present between the lobes (see exception below). There is an **oblique (major) fissure** on either side but there is only a **horizontal fissure** on the right so the side does not need to be specified in this case. The **lingula** is divided into **superior** and **inferior segments** (rather than **medial** and **lateral** in the **middle lobe**). In the left lung, the apical and posterior segments of the upper lobe, and anterior basal and medial basal segments of the lower lobe share a single tertiary bronchus and are thus known as the **apicoposterior** and **anteromedial segments**, respectively.

The secondary pulmonary lobule is an important concept in thoracic radiology. The bronchiole and pulmonary arteriole are in the centre and lymphatics and veins around the periphery in the interlobular septum. Note how the diameter of the arteriole is bigger than that of the adjacent bronchiole, in the absence of bronchiectasis. In normal conditions, the interlobular septum is too thin to be appreciable on imaging. Each secondary lobule supplies 10–12 acini. An acinus is the largest pulmonary unit where all airways take part in gas exchange.

The **inferior accessory fissure (Twining's line)** divides the medial basal segments from rest of the lower lobe.

Both Twining's line and the **azygos fissures** are anatomical variants.

Exam tips:
- The azygous fissure is the most commonly seen accessory fissure and is in the right upper lobe.
- Trivia: How many layers of pleura are there in an azygous fissure?
- Answer: Four (two visceral and two parietal).

Q2.18 PA frontal chest radiograph

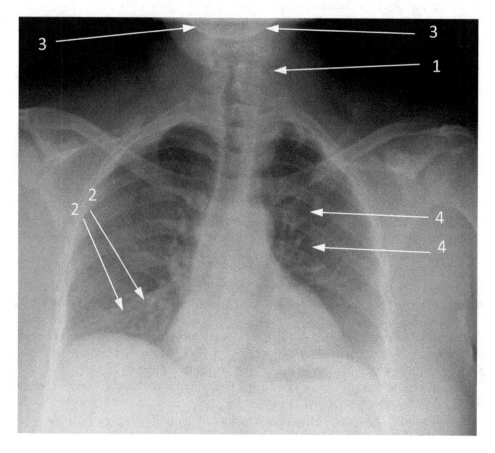

1. Name the arrowed structure.
2. Name the arrowed structure.
3. Name the arrowed structure.
4. Name the arrowed structure.
5. Name the anatomical variant.

Answers
1. Left pedicle of the C6 vertebra.
2. Bronchiole in the right lung.
3. Body of the mandible.
4. Medial border of the left scapula.
5. Left cervical rib.

Comments:
Cervical ribs are accessory ribs from the lowest (7th) cervical vertebra. They are usually bilateral (though not in this case) which can make their identification harder in an exam. Note the difference in curvature of the cervical rib compared to the first thoracic rib.

It isn't possible to be certain which lobe the **bronchiole** is located in on this projection, so it is enough to say that it is within the right lung.

Exam tip:
- The ability to label ribs and vertebral levels cannot be over-emphasised in preparing for the exam. It is a good idea to make sure that you are familiar with the key landmarks separating different parts of the spine.

Q2.19 PA frontal chest radiograph

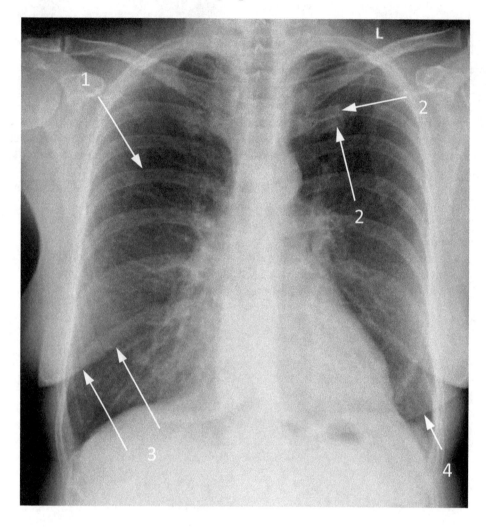

1. Name the arrowed structure.
2. Name the arrowed structure.
3. Name the arrowed structure.
4. Name the arrowed structure.
5. Name the anatomical variant.

Answers
1. Right 6th posterior rib.
2. Left 1st costochondral joint.
3. Skin crease.
4. Left breast shadow.
5. Pectus excavatum.

Comments:

The **costochondral joint** is where the rib attaches to the **costal cartilage**.

Skin creases and **breast shadows** are structures to be aware of as they are commonly seen on plain chest radiographs and can mimic lung pathology.

On a frontal radiograph, **pectus excavatum** can be identified by horizontally-oreintated posterior ribs, vertically-orientated anterior ribs, increased density of the inferomedial lung zone, and a blurred right heart border (all of which are demonstrated in this case). The sternal concavity is best appreciated on lateral radiographs.

Exam tip:
- The costochondral junction can mimic abnormal opacification in the lung. The key is to compare the two sides for symmetry.

Q2.20 Axial CT chest with contrast

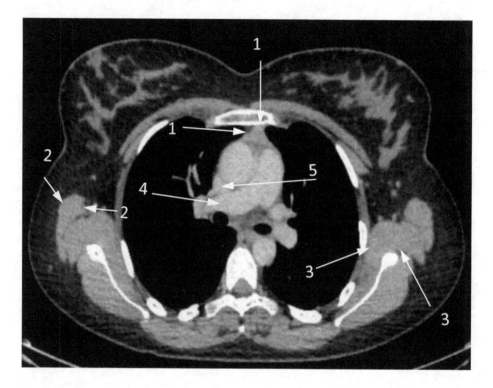

1. Name the arrowed structure.
2. Name the arrowed structure.
3. Name the arrowed structure.
4. Name the arrowed structure.
5. Name the arrowed structure.

Answers
1. Thymus.
2. Right latissimus dorsi muscle.
3. Left subscapularis muscle.
4. Right main pulmonary artery.
5. Superior vena cava.

Comments:
The **latissimus dorsi muscle** attaches the upper limb to the vertebral column and is the most lateral muscle through most of the thoracic region.

The **subscapularis muscle** lies medial and inferior to the **scapula** (hence its name). Together with **supraspinatus, infraspinatus** and **teres minor**, it forms the rotator cuff muscle group, which collectively acts to stablise the shoulder.

Exam tip:
- In addition to muscles in the limbs, it also important to learn the muscles of the torso, as they are less often considered but can be readily identified on cross-sectional imaging of the thorax or abdomen.

Q2.21 Coronal section from a CT pulmonary angiogram

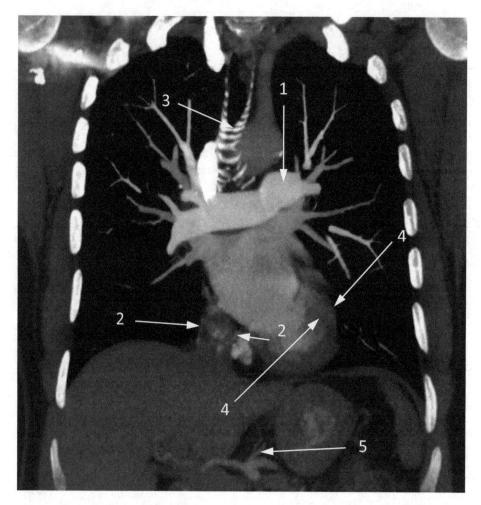

1. Name the arrowed structure.
2. Name the arrowed structure.
3. Name the arrowed structure.
4. Name the arrowed structure.
5. Name the arrowed structure.

Answers
1. Left main pulmonary artery.
2. Inferior vena cava.
3. A cartilaginous ring of the trachea.
4. Myocardium of the left ventricle.
5. Left gastric artery.

Comments:
The pulmonary trunk divides into two main **pulmonary arteries,** which then subdivide into lobar, segmental and subsegmental branches. Note the reflux of contrast in the **inferior vena cava,** which is a common finding on a CT pulmonary angiogram.

The trachea is supported by anterior **cartilagenous rings** and has a soft posterior membrane. The rings provide structural support and help keep the trachea rigid when it expands and lengthens during inspiration. In the expiratory phase, the posterior membraneous trachea becomes concave, which is best appreciated on axial slices.

The coeliac trunk has three main branches, the **splenic**, **left gastric** and **common hepatic arteries**.

Exam tip:
- The title states that this is a CT pulmonary angiogram, and so the contrast tracking is centred on the pulmonary arterial tree. An appreciation of the expected enhancement of structures during different contrast phases can help you to identify them during the exam.

Q2.22 Coronary CT angiogram

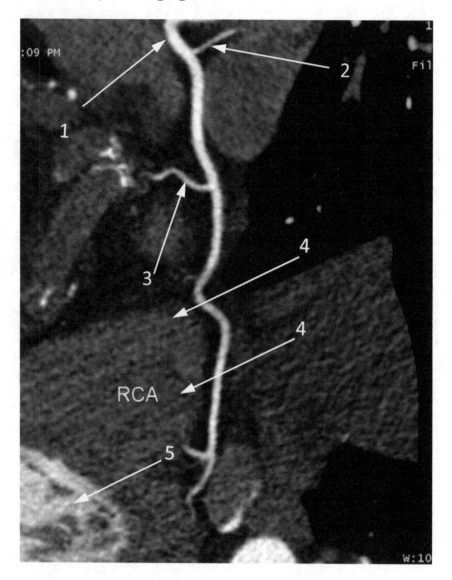

1. Name the arrowed structure.
2. Name the arrowed structure.
3. Name the arrowed structure.
4. Name the arrowed structure.
5. Name the arrowed structure.

Answers
1. Right coronary artery.
2. Conus branch of the right coronary artery.
3. Sinus node artery.
4. Right ventricle.
5. Left ventricle.

Comments:
The **right coronary artery** is labeled as 'RCA' on this image. Its first branch is the **conus branch** (in 50%–60% of the population), which supplies the right ventricle outflow tract. In the remainder of the population, the conus branch comes directly off the aorta.

The second branch of the RCA is the **sinus node artery** (in 60% of cases), which runs posteriorly to the sino-atrial node. In 40% of cases, the sinus node artery comes off the circumflex.

The clue to answering Q3 is given by the vessel supplying it, which is the acute marginal branch, so named because of the acute angle it makes with the RCA.

The **left ventricle** can be identified by the presence of papillary muscles within it.

Exam tips:
- For those unfamiliar with coronary CT angiograms, the imaged structures might look somewhat different to expected in the image given. You might be shown reconstructed images such as this one, where the structural orientation is adjusted to best illustrate structures of interests (coronary arteries in this case).
- The right coronary artery is labelled on the image in this question. It's important to pay attention to information that is given, especially on an ultrasound, angiographic or fluoroscopic images. This can make a seemingly difficult and obscure questions more accessible.

Q2.23 Sagittal thick-slab section from a contrast-enhanced CT of the chest

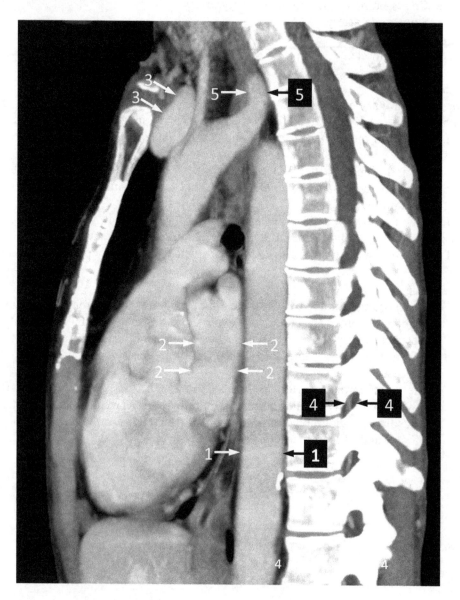

1. Name the arrowed structure.
2. Name the arrowed structure.
3. Name the arrowed structure.
4. Name the arrowed structure.
5. Name the arrowed normal variant.

Answers
1. Descending thoracic aorta.
2. Left atrium.
3. Left brachiocephalic vein.
4. Thoracic vertebral neural foramen.
5. Aberrant right subclavian artery.

Comments:

The **left atrium** lies between the **descending thoracic aorta** posteriorly and the **aortic root** anteriorly. It is a relatively midline structure in comparison to the right atrium, which lies more anteriorly and to the right. Notice the close relation of the **left main bronchus**, lying just superior to the left atrium on this image. When enlarged, the left atrium can displace the main bronchi superiorly causing splaying of the carina.

An **aberrant right subclavian** artery is a normal variant, but can be associated with dysphasia due to oesophageal compression (dysphasia lusoria). The conventional origin of the aortic arch vessels is for the (right) **brachiocephalic artery** to originate first, followed by the **left common carotid** (CCA) and finally the **left subclavian artery** (SCA). In the case of an aberrant right subclavian, the right common carotid artery originates first, but the right subclavian artery originates distal to the left subclavian artery, and often passes posterior to the oesophagus. This can give a distinctive appearance on barium swallows. Other variants of the aortic arch include a bovine arch (common origin of the brachiocephalic and left CCAs), variant origins of the vertebral arteries (normally branches of the SCAs), a right sided arch with aberrant left subclavian artery, or a double aortic arch.

Exam tip:
- Identifying this image as just left of the midline is important in determining the structures. Normally, the descending thoracic aorta in on the left and this is the biggest clue to the side. The other clue is the presence of the left ventricle, identified by its thick muscular wall, best seen posteriorly on this image.

Q2.24 Upper limb venogram

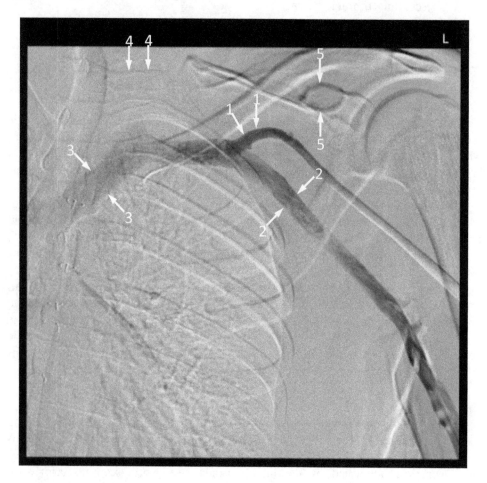

1. Name the arrowed structure.
2. Name the arrowed structure.
3. Name the arrowed structure.
4. Name the arrowed structure.
5. Name the arrowed structure.

Answers
1. Left cephalic arch.
2. Left axillary vein.
3. Left brachiocephalic vein.
4. Left transverse process of T1.
5. Coracoid process of the left scapula.

Comments:
The upper limb has three major draining veins: the **brachial** (often paired); **basilic** and **cephalic veins**. The basilic and brachial veins coalesce to form the **axillary vein** at level of the inferior margin of **teres major muscle**. At its distal aspect, the axillary vein is joined by the **cephalic vein** just before it crosses the lateral aspect of the first rib, at which point it becomes the **subclavian vein**. The subclavian vein coalesces with the **internal jugular vein** to form the **brachiocephalic vein**. The left and right brachiocephalic veins join to form the **superior vena cava**.

Exam tip:
- Identifying the lateral border of the first rib is an important anatomical landmark when defining whether the labelled vein is the axillary (lateral) or subclavian vein (medial). You might also have noticed the sternotomy wires on this image. In the exam, the images will not show any abnormality so these would not be present!

Q2.25 Coronal thick-slab section from a contrast-enhanced CT of the chest

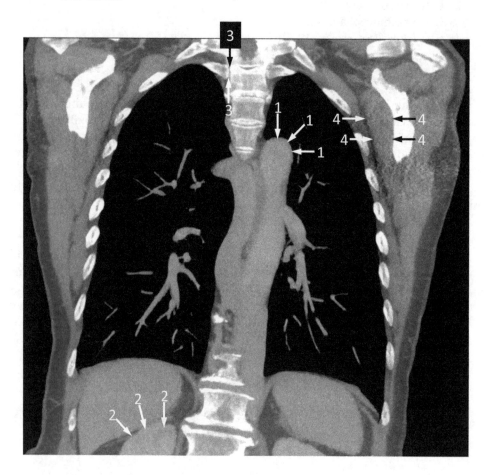

1. Name the arrowed structure.
2. Name the arrowed structure.
3. Name the arrowed structure.
4. Name the arrowed structure.
5. Name the anatomical variant.

Answers
1. Aortic arch.
2. Upper pole of the right kidney.
3. Right 2nd rib costovertebral joint.
4. Left subscapularis muscle.
5. Azygos/hemiazygos continuation of the inferior vena cava.

Comments:
Question 5 is particularly difficult. **Azygos continuation of the IVC** is a rare anatomical variant caused by absence of the supra-renal/infra-hepatic portion of the IVC. As a result, blood is redirected through a dilated **azygos vein**, which originates around the level of T12-L2, ascends the right side of the vertebral column before arching over the right main bronchus to join the **superior vena cava**. The **hemiazygos vein** originates at a similar level to the azygous on the left and ascends to the level of T8–9, where it crosses the midline to drain into the azygos vein. In this rare case, there is also a left-sided IVC (not seen on this image) and so the hemiazygos drains the infra-hepatic IVC, before anastomosing with the azygos, which can be seen on this image. Note, you can also see the more inferior portion of the (non-dilated) azygos just below the point where the hemiazygos joins.

The ribs articulate with the vertebrae at two locations: at the vertebral bodies (**costovertebral joints**) and the transverse processes (**costotransverse joints**).

Exam tip:
- Remember, chest CTs are normally performed with the arms above the patient's head (to reduce artefact). This has the result of rotating the scapulae, which is why the muscles associated with the scapula are well demonstrated on this coronal section.

IOP Publishing

Anatomy for the Royal College of Radiologists Fellowship
Illustrated questions and answers
Andrew G Murchison, Mitchell Chen, Thomas Frederick Barge, Shyamal Saujani, Christopher Sparks, Radoslaw Adam Rippel, Malcolm Sperrin and Ian Francis

Chapter 3

Abdomen and pelvis

Thomas Frederick Barge and Christopher Sparks

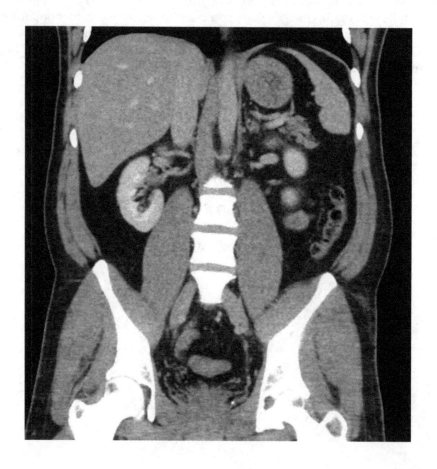

Q3.1 Oblique acquisition image from a barium swallow contrast study

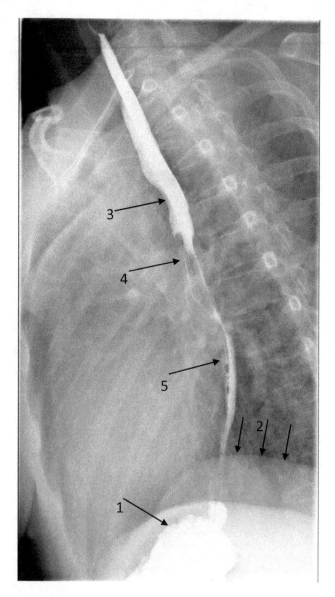

1. Name the arrowed structure.
2. Name the arrowed structure.
3. Which anatomical structure causes this impression on the oesophagus?
4. Which anatomical structure causes this impression on the oesophagus?
5. Which anatomical structure causes this impression on the oesophagus?

Answers
1. Stomach fundus.
2. Right hemidiaphragm.
3. Aortic arch.
4. Left main bronchus.
5. Left atrium.

Comments:
Interpretation of fluoroscopic images requires identification of the structure outlined by contrast, and understanding of nearby structures. The **oesophagus** itself has a simple anatomical path, but is adjacent to and impacted upon by three clinically important structures: the **arch of the aorta**, the **left main bronchus**, and the **right atrium**. Although the mediastinum is best illustrated on cross sectional imaging, you should ensure that you are confident of mediastinal anatomy in all imaging modalities.

The stomach is divided into:
- The **cardia**, containing the gastro-oesophageal junction
- The **fundus**, the most superior part of the stomach
- The **body**
- The **pylorus** (antrum), emptying into the small intestine.

Exam tips:
- The atypical view here is a good example of how examiners might try to challenge you—it has been taken obliquely in a supine patient, rather than the commonly used PA view.
- Due to the upper abdominal viscera and the heart, the **hemidiaphragms** are not at the same level. The left hemidiaphragm is usually one rib intercostal space (~2 cm) lower than the right.

Q3.2 PA acquisition image from barium follow through contrast study

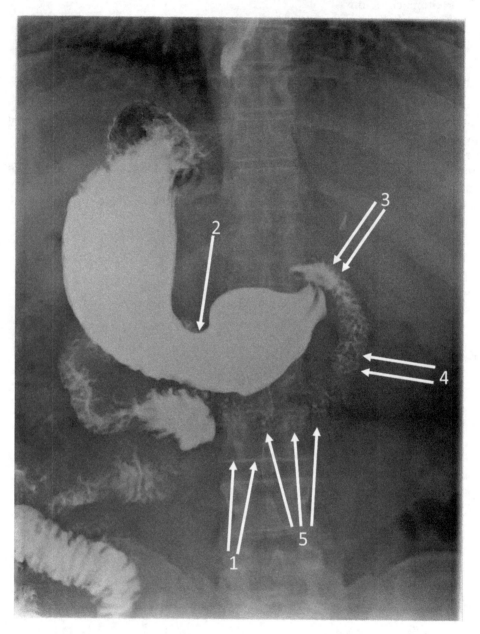

1. Name the arrowed structure.
2. Name the arrowed structure.
3. Name the arrowed structure.
4. Name the arrowed structure.
5. Name the arrowed structure.

Answers
1. Inferior endplate of L3 vertebra.
2. Incisura angularis.
3. D1/ First part of the duodenum.
4. D2/ Second part of the duodenum.
5. D3/ Third part of the duodenum.

Comments:
The duodenum is divided into four segments.

The **first part (D1, superior)** is continuous with the **pylorus**. It tracks superiorly and posteriorly towards the right, before turning inferiorly at the **superior duodenal flexure**. The first few centimetres of D1 are intraperitoneal, with the remainder of the duodenum being retroperitoneal.

Following this, the **second part (D2, descending)** courses caudally, wrapping around the **head of the pancreas**. Importantly, it contains the entrance of the **pancreatic duct** and **common bile duct** via the **major duodenal papilla (Ampulla of Vater)**, which is also an anatomical landmark between the embryological foregut and midgut.

The **third part (D3, inferior/horizontal)** begins at the **inferior duodenal flexure**, traversing horizontally to the left across the vertebral column (which helpfully acts as an anatomical knowledge landmark for candidates).

The **fourth part (D4, ascending)** is only partially demonstrated on this image. It courses cranially, either anterior to, or to the right of the aorta, before turning anteriorly and terminating at the **duodenojejunal (DJ) flexure**. The DJ flexure is a clinically significant landmark, as it is a fixed point where the suspensory muscle of the duodenum **(Ligament of Treitz)** attaches, and as it marks the start of the **jejunum**.

Exam tips:
- Familiarise yourself with the appearances of different sections of the small intestine. The **jejunum** has a feathery appearance, is located in the left upper quadrant of the abdomen, and has more **valvulae conniventes** and fewer folds per length than the relatively featureless ileum.

Q3.3 Axial slice from portal venous CT of the upper abdomen

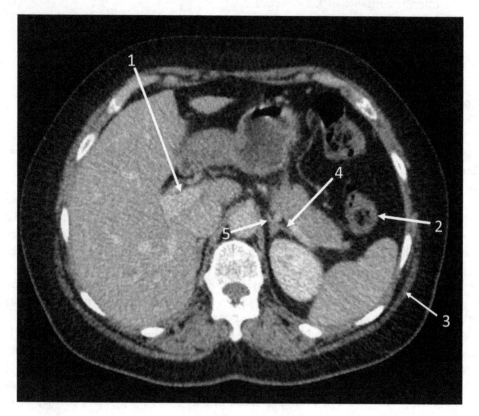

1. Name the arrowed structure.
2. Name the arrowed structure.
3. Name the arrowed structure.
4. Name the arrowed structure.
5. Name the arrowed structure.

Answers
1. Portal vein.
2. Descending colon.
3. Left latissimus dorsi muscle.
4. Lateral limb of the left adrenal.
5. Body of the left adrenal.

Comments:

The **portal vein** drains the gastrointestinal system and spleen, and originates behind the **neck of the pancreas** as the confluence of the **splenic vein** and the **superior mesenteric vein**. At the **porta hepatis** it divides into the **left and right portal veins**. It is useful to be aware of variant portal vein anatomy, the most common configuration being **portal vein trifurcation**.

The **adrenals** are paired endocrine glands which are asymmetric in shape. The left is slightly larger, with a triangular, Y-shaped appearance on axial imaging, in contrast to the linear, inverted V-shape of the right adrenal.

Each adrenal gland consists of a **body** anteriorly and two **limbs (lateral and medial)** extending posteriorly. The left adrenal gland lies anteromedial to the upper pole of the left kidney, and the right is located superior to the upper pole of the right kidney, between the right **crus of the diaphragm** and **right lobe of the liver**.

Exam tip:
- It can sometimes be difficult to distinguish the **splenic vein** from the **portal vein** on a single image. The only way to definitively differentiate the two is by identifying the **portovenous confluence** of the superior mesenteric vein (which courses superiorly) and the splenic vein. This is well demonstrated on angiography and coronal imaging, but not readily visible on single slice axial images. If in doubt, recall that the portal vein originates **behind the neck of the pancreas**—if arrows are positioned to the right to the right of this the answer is likely to be the portal vein.

Q3.4 Coronal slice from portal venous CT abdomen/pelvis

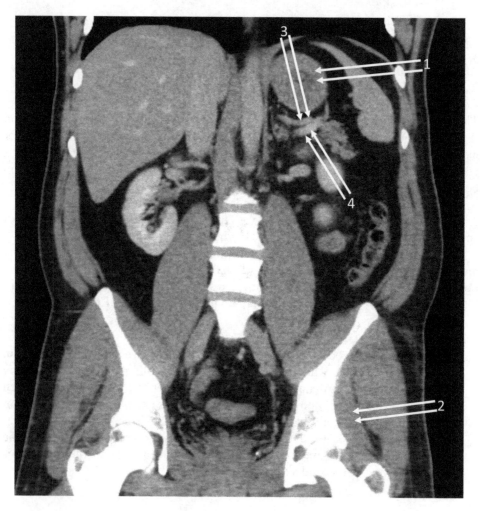

1. Name the arrowed structure.
2. Name the arrowed structure.
3. Name the arrowed structure.
4. Name the arrowed structure.
5. Name the anatomical variant.

Answers
1. Stomach.
2. Left gluteus minimus muscle.
3. Splenic artery.
4. Splenic vein.
5. Right accessory renal artery/right aberrant renal artery.

Comments:

Questions with coronal images such as this can be difficult, testing knowledge of the relationships between different structures.

The **splenic artery**, the largest of the three branches of the **coeliac artery**, traverses superiorly along the **pancreas**. It is differentiated from the **splenic vein**, both in its serpiginous appearance (compared to the relatively straight splenic vein) and its anterosuperior position. At the **splenic hilum** it divides into anterior and superior terminal branches, with each of these further dividing into 4–6 segmental arteries. A common anatomical variant is for the splenic artery to arise directly from the **abdominal aorta**.

Accessory renal arteries are another common anatomical variant. They can be unilateral or bilateral.

Exam tip:
- **Accessory renal arteries** most commonly accessory arise from the abdominal aorta; however, they may also arise from the coeliac, superior/inferior mesenteric, middle colic or middle sacral arteries.

Q3.5 PA acquisition image from endoscopic retrograde cholangio—pancreatography (ERCP)

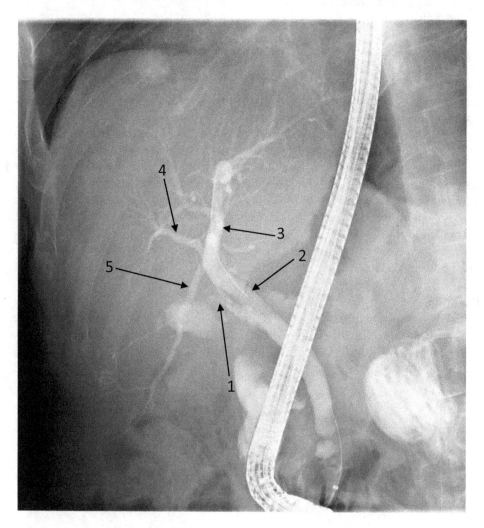

1. Name the arrowed structure.
2. Name the arrowed structure.
3. Name the arrowed structure.
4. Name the arrowed structure.
5. Name the arrowed structure.

Answers
1. Cystic duct.
2. Common hepatic duct.
3. Left hepatic duct.
4. Right posterior duct.
5. Right anterior duct.

Comments:
The biliary tree is conventionally separated into intra- and extra-hepatic bile ducts. The intrahepatic segmental bile ducts combine to form sectional bile ducts. These are:
- The **left hepatic duct**, draining segments II-IV
- The **right anterior duct**, draining the right lobe medial segments V and VIII
- The **right posterior duct**, draining the right lobe lateral segments VI and VII.

The right anterior and posterior ducts combine to form the **right hepatic duct** which, together with the **left hepatic duct**, forms the **common hepatic duct**.

The extrahepatic ducts consist of the common hepatic duct and **cystic duct** (arising from the **gallbladder**), which combine to form the **common bile duct**. Posterior to the duodenum and pancreas the common bile duct combines with the **main pancreatic duct** to form the **hepatopancreatic ampulla (ampulla of Vater)**, draining into the **second part (D2) of the duodenum**.

Exam tip:
- Anatomical variation in bile duct anatomy is common, making intrahepatic sectional duct anatomy a less likely topic in the exam.

Q3.6 Coronal reformat from a CT Urogram

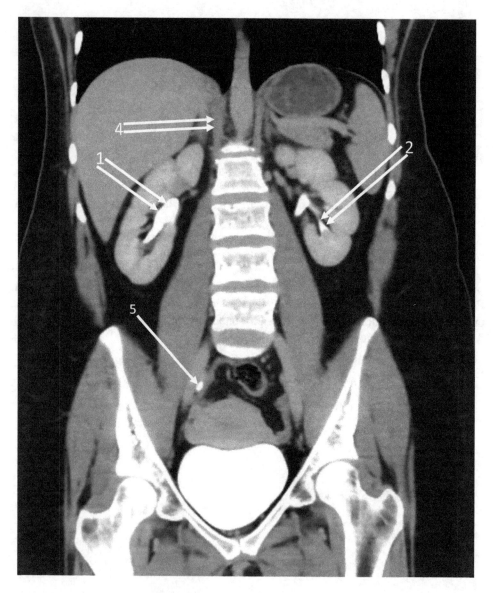

1. Name the arrowed structure.
2. Name the arrowed structure.
3. Name the normal variant on this image.
4. Name the arrowed structure.
5. Name the arrowed structure.

Answers

1. Right renal pelvis.
2. Left lower pole minor calyx.
3. Persistent foetal lobulation of the kidneys.
4. Right crus of diaphragm.
5. Right ureter.

Comments:

Careful inspection of the left lower pole allows us to demarcate a cone shaped **renal pyramid**, with its base facing the renal cortex, and its apex forming a papilla which excretes urine into a **minor calyx**. The acute angle formed at the corner of the papilla and minor calyx is known as the fornix, which become blunted in the setting of hydronephrosis. Two or three minor calyces converge to corm a **major calyx**, which in turn join to form the **renal pelvis**.

The **diaphragmatic crura** blend with the **anterior longitudinal ligament** of the vertebral column, attaching to the anterior bodies of the lumbar vertebra. The right crux is slightly longer than the left, and arises from L1–3 compared to L1–2 on the left.

Note the lobulated shape of the kidneys in this image, in particular the left kidney. Persistent foetal lobulation is a normal variant seen in adults as a result of incomplete fusion of the developing renal lobules. It is seen as a smooth lobulation of the kidney outline, and can be demonstrated on CT, US or MRI.

Exam tip:

- Another anatomical variant to be aware of is the dromedary hump. This is a prominent focal bulge on the lateral aspect of the left kidney, caused by the splenic impression. Like foetal lobulation, it can be seen on CT, US or MRI.

Q3.7 Portal venous phase, axial slice of a CT of the upper abdomen

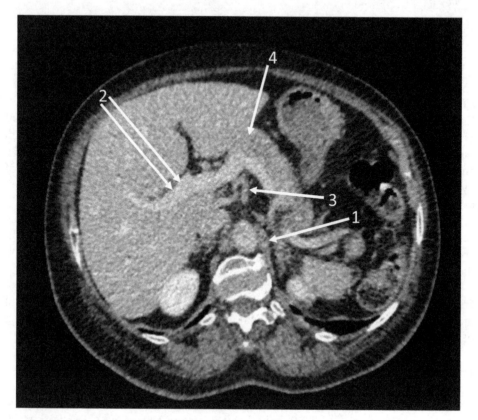

1. Name the arrowed structure.
2. Name the arrowed structure.
3. Name the arrowed structure.
4. Name the arrowed structure.
5. Name the anatomical variant.

Answers
1. Left crus of diaphragm.
2. Right portal vein.
3. Superior mesenteric artery.
4. Neck of pancreas.
5. Right hepatic artery arising from superior mesenteric artery.

Comments:
The portal vein originates at the portovenous confluence, posterior to the **pancreatic neck** and just to the right of the midline. It courses superiorly within the **hepatoduodenal ligament**, dividing at the **porta hepatis** into the **right and left portal veins**.

The **right portal vein** divides into anterior and posterior branches, the anterior of which is demonstrated on this image. The **left portal vein** gives off branches to **segments I and IV** (Caudate and Quadrate lobes), prior to dividing into superior and inferior branches more distally.

The **neck of the pancreas** is the thinnest part, and is the only strictly anatomically defined part of the pancreas, lying anteriorly to the portovenous confluence and **superior mesenteric artery**. The **head of the pancreas** is positioned to the right of the superior mesenteric vessels, with the **body** to the left of them. The **tail** lies between the layers of the splenorenal ligament in the **splenic hilum**.

Exam tips:
- Variation in **hepatic artery** anatomy is very common, seen in approximately 40% of the population.
- The two most common variants are a **replaced right hepatic artery** from the superior mesenteric artery, and a **replaced left hepatic artery** from the left gastric artery.

Q3.8 Axial slice from portal venous phase CT abdomen

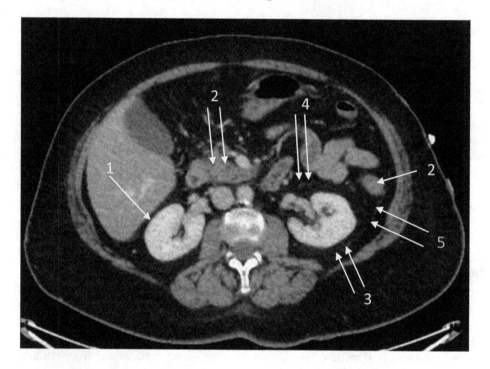

1. Name the compartment that this structure is contained in.
2. Name the compartment that these structures are contained in.
3. Name the arrowed structure.
4. Name the arrowed structure.
5. Name the arrowed structure.

Answers
1. Right perirenal space.
2. Anterior pararenal space.
3. Left posterior renal fascia (Zuckerkandl's fascia).
4. Left anterior renal fascia (Gerota's fascia).
5. Left lateral conal fascia.

Comments:
There are three compartments of the retroperitoneum: the anterior pararenal space, the perirenal space and the posterior pararenal space. These are separated by the **anterior (Gerota's)** and **posterior (Zuckerkandl's) renal fascia**, and the **lateral conal fascia**. Note that the anatomical (rather than eponymous) names provide an aide memoire for the anatomy, and avoids unnecessary dropping of marks by mistaking one eponymous structure for another.

The contents of the retroperitoneal compartments are listed below:

Anterior pararenal space:
- Ascending colon
- Descending colon
- 2nd and 3rd parts of duodenum
- Pancreas

Perirenal space (surrounds each kidney):
- Kidneys
- Adrenals
- Proximal ureter

Posterior pararenal space:
- Fat
- Strictly a 'potential space'—important clinically as an area for potential disease spread.

Exam tips:
- Although retroperitoneal anatomy may seem esoteric, it could legitimately come up in the exam, and a little knowledge goes a long way.
- It might be helpful to take a few moments to look at diagrams of retroperitoneal compartments online.

Q3.9 Coronal T2-weighted slice from a MRI examination of the liver

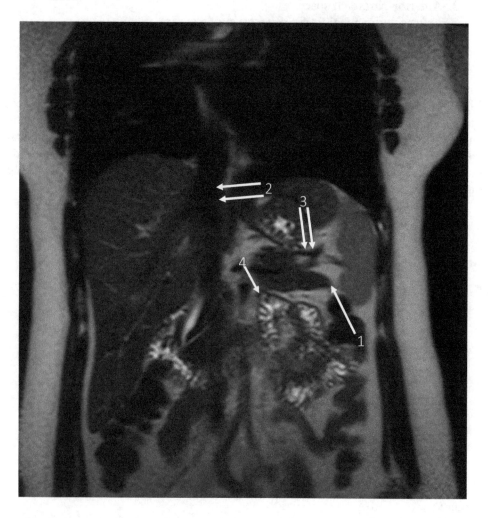

1. Name the arrowed structure.
2. Name the arrowed structure.
3. Name the arrowed structure.
4. Name the arrowed structure.
5. Name the anatomical variant.

Answers
1. Pancreatic tail.
2. Inferior vena cava.
3. Splenic artery.
4. Duodenojejunal flexure.
5. Riedel lobe.

Comments:

A **Riedel lobe** is a very common anatomical variant, more frequently found in females than males. It is defined as a tongue-like inferior projection of the right lobe of the liver, beyond the level of the most inferior costal cartilage.

Note the parts of the **duodenum** demonstrated on this coronal image. The **jejunum** typically lies within the left upper quadrant, and has a relatively delicate, feathery appearance.

As discussed in previous questions, the **splenic artery** courses superiorly to the **splenic vein**, taking a tortuous course. If you answered this incorrectly, please revisit question four.

Exam tip:
- If in doubt about a structure, always consider adjacent structures. You may, for example, be uncertain about question 1, but its position adjacent to the splenic hilum confirms that it is the pancreatic tail.

Q3.10 Lateral acquisition from a barium swallow examination

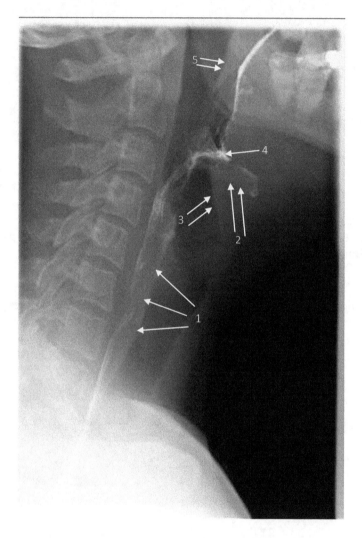

1. Name the arrowed structure.
2. Name the arrowed structure.
3. Name the arrowed structure.
4. Name the arrowed structure.
5. Name the arrowed structure.

Answers
1. Oesophagus.
2. Hyoid bone.
3. Laryngeal vestibule.
4. Vallecula.
5. Soft palate.

Comments:
The **epiglottis** is a mucosal flap attached at the base of the thyroid cartilage, projecting dorsally towards the base of the tongue. Upon swallowing, it shifts inferiorly to cover the airway, directing food towards the **piriform sinuses (piriform fossae)** at either side, preventing aspiration. The epiglottic tubercle, a prominence at the lower part of the epiglottis above the stalk of the epiglottis, may be seen at the anterior wall of the **laryngeal vestibule**. The laryngeal vestibule lies between the laryngeal inlet and the vocal folds (true vocal cords) below, and contains the vestibular folds (false vocal cord) and laryngeal ventricle (Morgagni's sinus).

The **valleculae** are paired recesses within the glossoepiglottic folds, extending from the base of the tongue to the anterior epiglottis. These act as 'spit traps', holding saliva temporarily to prevent the initiation of the swallow reflex. The valleculae and piriform sinuses can be demonstrated on barium swallow examination, both in anteroposterior and lateral views. They are important anatomical landmarks for the exam and clinical practice.

Exam tip:
- The **valleculae** and **piriform sinuses** are easy to confuse—remember the pIrIform sinuses lie Inferior to the valleculae.

Q3.11 Transverse view from an ultrasound of the upper abdomen

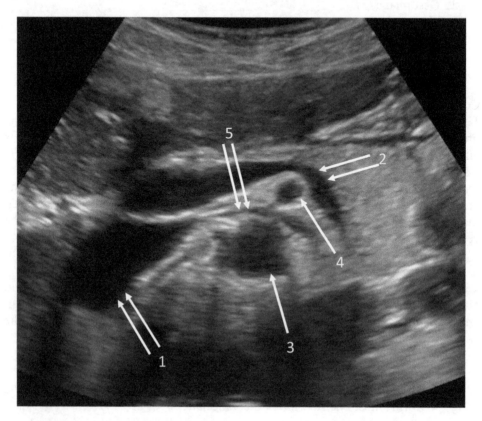

1. Name the arrowed structure.
2. Name the arrowed structure.
3. Name the arrowed structure.
4. Name the arrowed structure.
5. Name the arrowed structure.

Answers
1. Inferior vena cava.
2. Splenic vein.
3. Abdominal aorta.
4. Superior mesenteric artery.
5. Left renal vein.

Comments:
Of all the imaging modalities, ultrasound probably provides the greatest test of knowledge of the relationship of anatomical structures to one another.

The image in question is the standard transverse view of the upper abdomen, often obtained at the beginning of abdominal ultrasound examinations. The **pancreas** can be seen running leftward and posteriorly, curving towards the **spleen**. Note the slightly hyperechoic appearances—in children the pancreas contains significantly less fat, and consequently can appear hypoechoic; this needs to be considered when answering questions involving the paediatric abdomen.

The **splenic vein**, coursing the superior border of the pancreas, can be seen forming the **portal vein** at the **portovenous** confluence (the **superior mesenteric vein** is not demonstrated on this image). Posterior to this, the superior mesenteric artery (SMA) descends inferiorly. The **abdominal aorta** is more posterior still, lying just in front of the **vertebral bodies**.

The **inferior vena cava** can be seen posterior to the portal vein, with the **left renal vein** draining into it. Note the position of the left renal vein, running between the aorta and SMA.

Exam tips:
- Try to spend some time in ultrasound prior to the exam—this will allow you to familiarize yourself with the anatomical appearances.
- Always remember to orientate yourself with the help of the title of the image—this is particularly useful in ultrasound images, with the limited field of view.

Q3.12 Transverse image from ultrasound of the upper abdomen

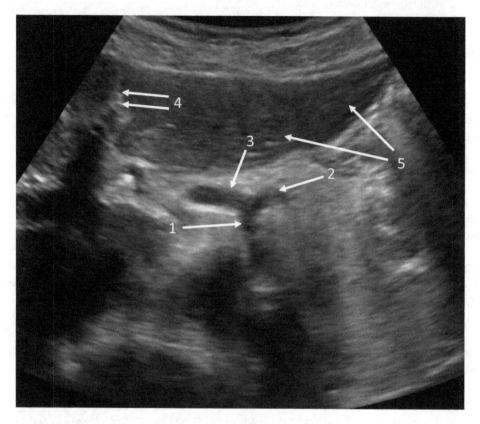

1. From behind which ligament does this structure arise?
2. Name the arrowed structure.
3. Name the arrowed structure.
4. Name the arrowed structure.
5. Name the arrowed structure.

Answers
1. Median arcuate ligament.
2. Splenic artery.
3. Common hepatic artery.
4. Falciform ligament.
5. Left lobe of the liver.

Comments:

The first major branch of the **abdominal aorta**, the **coeliac artery**, arises around the level of T12, behind the median arcuate ligament. This ligament is formed of the left and right crus of the diaphragm and has clinical relevance in median arcuate ligament syndrome (coeliac artery compression syndrome), where it compresses the coeliac trunk.

In the normal anatomical configuration (70% of the population) the small **left gastric artery** branches from the coeliac trunk first, with the **splenic artery** and common **hepatic artery** bifurcating more distally. Approximately 30% of the population have variant coeliac artery anatomy; any of the three main branches may take origin from the aorta or superior mesenteric arteries, and other arteries can arise directly from the coeliac artery (e.g. right hepatic, gastroduodenal arteries).

Exam tip:
- The common hepatic and splenic arteries bifurcate from the coeliac axis giving the characteristic ultrasound 'whale tail' appearance on transverse views of the upper abdomen.

Q3.13 AP plain abdominal radiograph

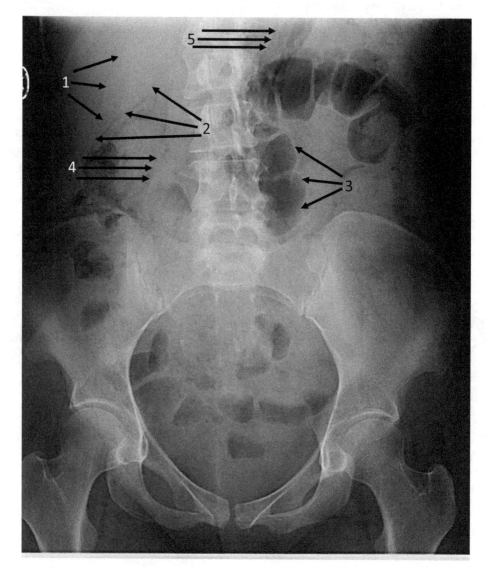

1. Name the arrowed structure.
2. Name the arrowed structure.
3. Name the arrowed structure.
4. Name the arrowed structure.
5. Name the arrowed structure.

Answers
 1. Right kidney.
 2. Liver edge.
 3. Transverse colon.
 4. Right psoas muscle.
 5. Gastric rugae.

Comments:

The differing densities of structures allows them to be differentiated from one another on plain radiographs. The bowel is easily demonstrated due to luminal gas, which is of low attenuation and therefore provides excellent contrast. More subtly, the abdominal viscera can be outlined.

The kidneys are surrounded by perinephric fat, and can be identified on most abdominal radiographs, the **upper pole** projected over the lower ribs and the **lower pole** adjacent to the **psoas muscle**, as demonstrated here. The liver can be seen as a relatively homogeneous opacity in the right upper quadrant, with the liver edge being appreciable in this image as it abuts the large bowel. The spleen is harder to visualise due to its smaller size and position, with only the inferior border extending beyond the lower ribs. On closer inspection of the image the inferior edge can just about be demarcated, the **splenic flexure of the colon** overlying it.

The rugal fold pattern of the **stomach** changes with peristaltic activity. At the **lesser curve**, the **rugae** are arranged longitudinally in parallel rows, known as Magenstrasse (main street). In other areas the rugae have no fixed pattern. Within the **antrum of the stomach** the muscosa contains 'area gastricae'. These are small undulations, giving the appearance of 2–3 mm 'cobblestones' which, on contrast studies, result in a mosaic pattern.

Exam tips:

 • Spend a little time studying plain abdominal radiographs on a DICOM viewer in your department—you might be surprised about how much can be seen.
 • Remember that the exam is digital, and as a result images can be manipulated by zooming and windowing, a useful feature for subtle questions such as this one. However, we recommend you use this with caution, as images in the exam are presented in a way that is optimised for viewing the arrowed structure (although you can always reset the image if your windowing gets out of control!)

Q3.14 AP abdominal radiograph contrast follow through study

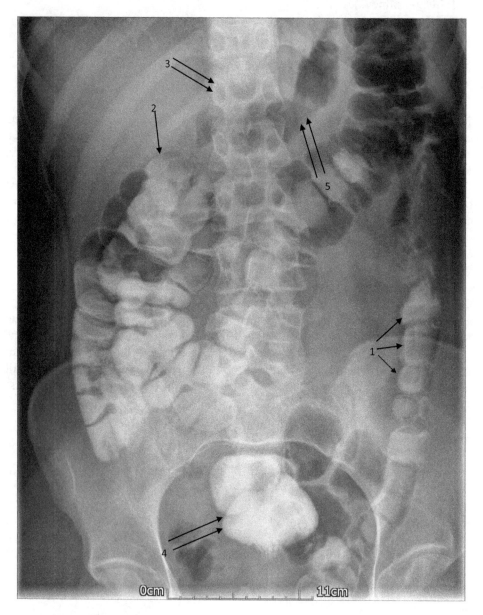

1. Name the arrowed structure.
2. Name the arrowed structure.
3. Name the arrowed structure.
4. Name the arrowed structure.
5. Name the arrowed structure.

Answers
1. Descending colon.
2. Hepatic flexure.
3. Right pedicle of T12 vertebra.
4. Rectum.
5. Greater curvature of stomach.

Comments:
Key imaging features that differentiate small bowel from large bowel on abdominal radiographs (contrast and non-contrast):

	Small bowel	Large bowel
Position	Central	Peripheral ('framing')
Septations	Complete	Incomplete
	Valvulae conniventes	Haustra
Diameter	<3 cm	<5.5 cm
Content	Fluid	Faeces

In the proximal colon the haustral sacs are fixed structures; more distally they are formed by muscular contractions, a feature well illustrated in the descending colon of the image on the previous page. Collapsed colon in contrast studies can have linear mucosal pattern, due to its longitudinal muscular bands (taenia coli).

Exam tip:
- The ascending and descending colon are extra-peritoneal structures, contained within the anterior pararenal spaces.

Q3.15 Sagittal slice from portal venous CT abdomen/pelvis

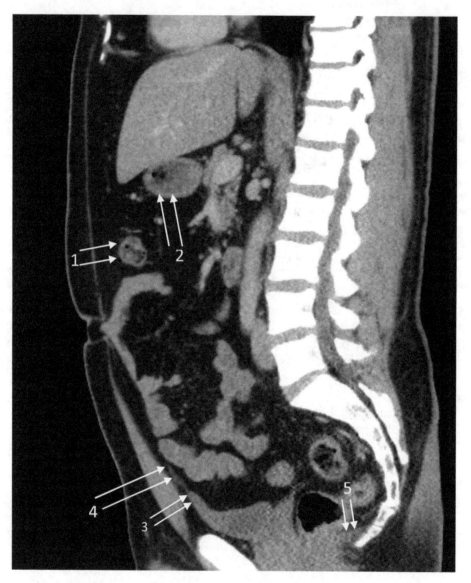

1. Name the arrowed structure.
2. Name the arrowed structure.
3. Name the arrowed structure.
4. Name the arrowed structure.
5. Name the arrowed structure.

Answers

1. Transverse colon.
2. Stomach.
3. Apex of urinary bladder.
4. Median umbilical ligament.
5. Pubococcygeus muscle.

Comments:

The **urinary bladder** has a triangular shape, becoming roughly oval when full. Anatomically it is divided into four regions: the **fundus** (or base) posteriorly, the **apex** anteriorly, the **body** between the apex and fundus, and the **neck** inferiorly, formed by the convergence of the two inferolateral surfaces and the fundus. The **ureters** enter the bladder inferoposteriorly on each side, and, along with the **urethra**, form the smooth-walled **trigone** within the fundus.

The superior aspect of the bladder is commonly known as the dome of the bladder, and is covered by a peritoneal reflection. Anteriorly at the apex, the **median umbilical ligament** attaches from the umbilicus, and is a remnant of the embryonic urachus.

The **levator ani** is a musculotendinous sheet, forming the majority of the floor of the pelvis. It consists of three components:

- **Puborectalis**—a U-shaped sling, extending from the **pubic symphysis** to the anorectal junction posteriorly.
- **Pubococcygeus**—running posteromedially from the bodies of the pubic bones to the **coccyx** and anococcygeal ligament.
- **Iliococcygeus**—starting anteriorly at the ischial spines and inserting posteriorly at the coccyx and anococcygeal ligament.

Exam tip:

- Familiarise yourself with sagittal and coronal views using the thin-slice reformats—mentally testing yourself as you scroll through, then confirming your answers on the cross-referenced axial images. As you do this, note the differences in appearance, density and enhancement (depending upon contrast protocol) of the structures. Keep repeating the task until you are consistently answering correctly.

Q3.16 Defaecating proctogram

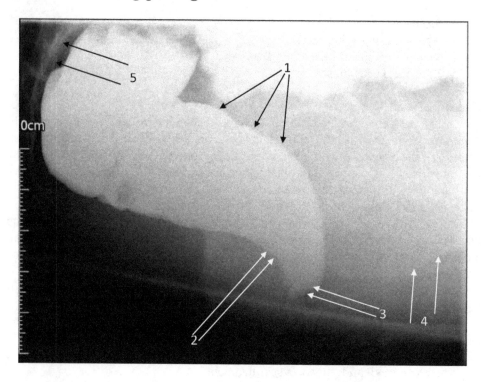

1. Name the arrowed structure.
2. What anatomical structure causes this impression?
3. Name the arrowed structure.
4. Name the arrowed structure.
5. Name the arrowed structure.

Answers
1. Rectum.
2. Puborectalis muscle.
3. Anorectal junction.
4. Femur.
5. Sacrum.

Comments:
Defaecating proctograms allow the assessment of anorectal function during defaecation. The anatomy is relatively simple if you've seen it before.

The examination is performed retrograde by filling the rectum with barium mixed with a thickening agent. The patient is then asked to defaecate under fluoroscopy, allowing functional assessment of the pelvic floor.

The **levator ani** forms the majority of the musculotendinous floor of the pelvis, providing support for the pelvic organs as well as aiding in faecal and urinary evacuation and continence. It is comprised of three muscles: the pubococcygeus, iliococcygeus and puborectalis muscles.

The **puborectalis muscle** is the most important muscle in regards to faecal continence, wrapping around the rectum in a horseshoe pattern with its insertion anteriorly at the **symphysis pubis**. When contracted it pulls the rectum tight, angulating it to prevent defaecation. This leaves an angle upon the outline of the rectum, as seen in in the question image. Upon puborectalis relaxation the angle is straightened, allowing evacuation. The **anorectal junction** is the right angle the rectum takes at the levator ani, transitioning to the **anal canal** thereafter.

Exam tip:
- Make sure you spend time revising a variety of imaging modalities. The college specifies that: 'in each exam roughly one-third of questions will be on images from cross-sectional techniques, plain radiographs and contrast studies (including those acquired by cross sectional means)'. This question is another example of where a little knowledge, or even better, experience, goes a long way.

Q3.17 Axial slice from a portal venous phase CT abdomen

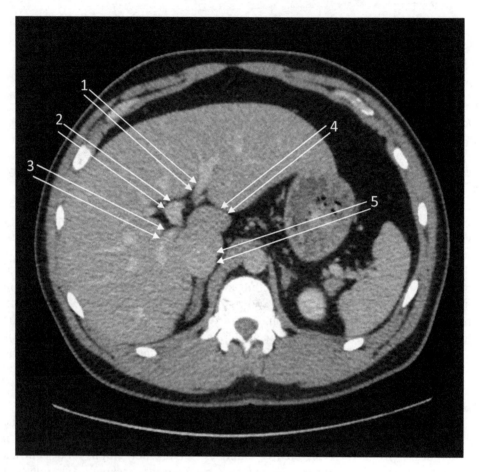

1. Name the arrowed structure.
2. Name the anatomical variant.
3. Name the anatomical variant.
4. Name the arrowed structure.
5. Name the arrowed structure.

Answers

1. Left portal vein.
2. Right anterior portal vein.
3. Right posterior portal vein.
4. Caudate lobe/hepatic segment I.
5. Inferior vena cava.

Comments:

Portal vein variant anatomy is seen in approximately 25% of the population and is important to recognise prior to procedures. **Trifurcation of the portal vein** is the most common variant, with the right portal vein bifurcating into anterior and posterior branches, as demonstrated in the image.

Rarer variants include portal vein duplication, an absent right portal vein and an absent left extrahepatic portal vein.

The **caudate lobe** is located posteriorly within the liver. It is anatomically different from the other liver segments, in that it has separate hepatic veins which connect directly to the **inferior vena cava**. Portal vein supply to the caudate lobe may be by both the left and right portal veins.

Exam tip:

- If you are really lost, it can be helpful to identify anatomical structures you are confident about that are local to the arrowed structure. By familiarising yourself with the anatomy immediately surrounding it, you may be able to deduce the answer.

Q3.18 Axial slice image from portal venous CT abdomen

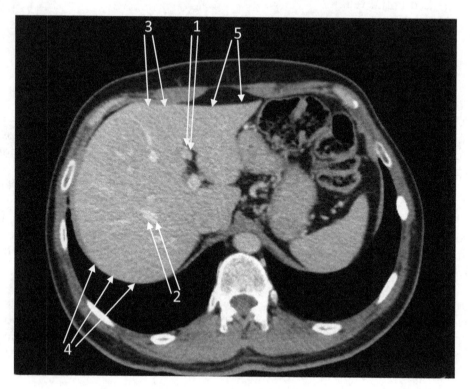

1. Name the arrowed structure.
2. Name the arrowed structure.
3. Name the hepatic segment arrowed.
4. Name the hepatic segment arrowed.
5. Name the hepatic segment arrowed.

Answers
1. Left hepatic vein.
2. Right hepatic vein.
3. Hepatic segment IV/IVa.
4. Hepatic segment VII.
5. Hepatic segment II.

Comments:
The Couinaud classification is the most widely accepted system to describe liver anatomy, dividing the liver into eight functional units (segments).

The vertically running **hepatic veins** separate the liver into four sections axially; the **right hepatic vein** divides the right lobe into right lateral and medial sections, the **middle hepatic vein** divides the liver into right and left lobes, and the **left hepatic vein** divides the left lobe into left medial and left lateral sections.

The sections are further divided horizontally into superior and inferior segments by the **right** and **left portal veins**. Segments are numbered in a clockwise fashion starting in the left lobe of the liver:
- Left lateral section: segment II superiorly and III inferiorly.
- Left medial section: segment IV, further divided into IVa superiorly and IVb inferiorly.
- Right medial section: segment VIII superiorly and V inferiorly.
- Right lateral section: segment VII superiorly and VI inferiorly.
- Note that segment I is the **caudate lobe.**

Exam tips:
- Segment IV is remembered by **IVa A**bove and **IVb B**elow.
- It is recommended that you review diagrams and spend time looking at cross sectional imaging (including reformats) to familiarise yourself with hepatic segmental anatomy. Remember that this can be asked on CT/MRI and on ultrasound studies.

Q3.19 Abdominal radiograph

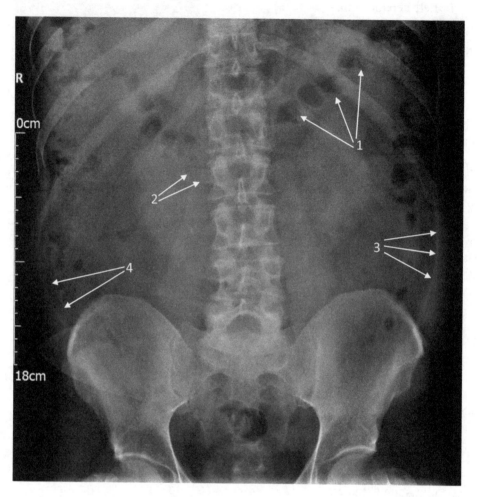

1. Name the arrowed structure.
2. Name the arrowed structure.
3. Name the arrowed structure.
4. Name the arrowed structure.
5. Name the anatomical variant.

Answers

1. Transverse colon.
2. Right transverse process of L2.
3. Left transversus abdominis muscle.
4. Right properitoneal fat stripe.
5. Horseshoe kidney.

Comments:

The **horseshoe kidney** is the most common congenital fusion variant, with the majority being fused at the lower poles and connected via an isthmus of functioning renal parenchyma or fibrous tissue. The normal ascent of the kidneys is prevented by the **inferior mesenteric artery** hooking over the isthmus, leaving the horseshoe kidney in a low-lying position. The **ureters** exit bilaterally and pass anteriorly over the isthmus.

The **properitoneal fat** is deep to the transversalis fascia, filling the posterior pararenal space. On imaging it can be seen as a line or stripe, and can be mistaken for pneumoperitoneum or pneumoretroperitoneum. Loss of the stripe is associated with pathology, including appendicitis and ruptured abdominal aortic aneurysm.

Exam tip:

• Spend an extra moment on each question ensuring you know exactly what the arrow is pointing at. If it is directed towards a specific part of anatomy (such as the abdominal wall muscle layer of transversus abdominis in this question) then you will need to provide a precise answer to obtain full marks.

Q3.20 Coronal slice from portal venous CT abdomen

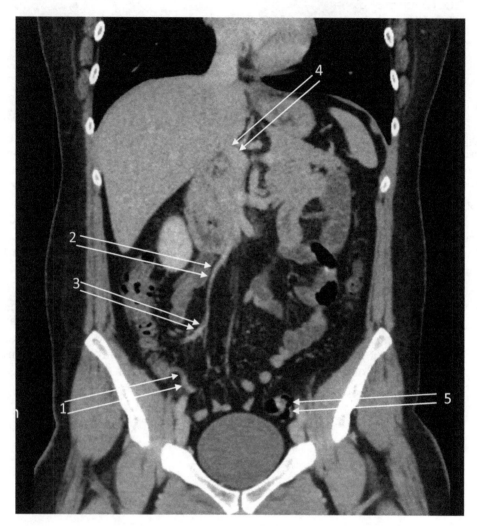

1. Name the arrowed structure.
2. Name the arrowed structure.
3. Name the arrowed structure.
4. Name the arrowed structure.
5. Name the arrowed structure.

Answers
1. Appendix.
2. Superior mesenteric vein.
3. Ileocolic vein.
4. Portal vein.
5. Sigmoid colon.

Comments:
The **appendix** is a 2–20 cm long blind muscular tube, arising from the posteromedial aspect of the **caecum**. In cross-sectional imaging it is most easily visualised in coronal slices, 2–3 cm inferior to the **ileocaecal valve** which is identifiable due to its fatty appearance (with low attenuation on CT). The tip of the appendix can have a highly variable position, being able to rotate almost 360 degrees, with a retrocaecal location being most common.

Exam tip:
• A duplex appendix is a rarely seen anatomical variant.

Q3.21 Digital subtraction angiogram

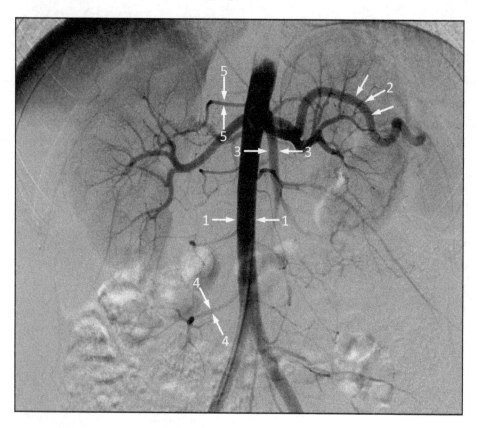

1. Name the arrowed structure.
2. Name the arrowed structure.
3. Name the arrowed structure.
4. Name the arrowed structure.
5. Name the arrowed normal variant.

Answers
1. Abdominal aorta.
2. Splenic artery.
3. Superior mesenteric artery.
4. Ileocolic artery.
5. Accessory right renal artery (supplying part of the right upper pole).

Comments:
The major branches of the abdominal aorta from cranial to caudal are: **Coeliac axis**, **superior mesenteric artery**, **renal arteries** and the **inferior mesenteric artery**. There are multiple lumbar branches, but these are often more difficult to define clearly on angiograms. The **superior mesenteric artery** has a characteristic descending appearance, which is how it can be identified here. It has multiple branches, terminating in the **ileocolic artery** which supplies the terminal ileum and caecum. The **coeliac axis** splits early, typically into 3 main branches, the **splenic** (with its characteristic corkscrew appearance), the **common hepatic** and the **left gastric** arteries. The **inferior mesnteric artery** is the smallest of the mesenteric branches and supplies the colon from the splenic flexure to the rectum. The aortic bifurcation is normally at the level of L1.

Exam tips:
- Digital subtraction angiography is a common way to define vascular anatomy. In this case a pigtail catheter has been placed in the abdominal aorta at the level of the renal arteries. It is highly likely that an angiogram or venogram will come up in the exam. Your initial thought should be to decide if the vasculature depicted is arterial or venous, as on first inspection, they may look quite similar.
- There are several clues on this example that this is an angiogram: 1. The main vessel (the aorta) lies slightly to the left (closer to the left kidney). 2. The renal arteries are opacified. If this was a venogram, no contrast would enter the renal veins as the flow of blood is in the wrong direction. 3. The title of the question!

Q3.22 Micturating cystourethrogram

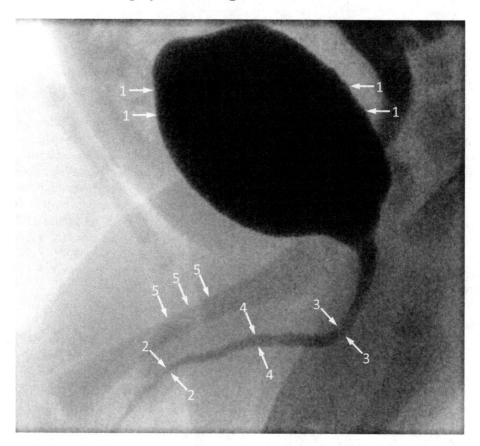

1. Name the arrowed structure.
2. Name the arrowed structure.
3. Name the arrowed structure.
4. Name the arrowed structure.
5. Name the arrowed structure.

Answers
1. Urinary bladder.
2. Penile urethra.
3. Membranous urethra.
4. Bulbous urethra.
5. Femoral shaft/diaphysis.

Comments:
The male urethra is divided into four parts: **prostatic, membranous, bulbous and penile.** The **membranous urethra** is the shortest and narrowest part of the urethra. It lies between the apex (bottom) of the prostate and the bulbous urethra and is recognised on this image by its antero-inferior course and slight concave shape. It is important to know about as it is the most common site of a traumatic urethral injury, because it is relatively fixed as it traverses the urogenital membrane.

Exam tips:
- It is important to inspect the image carefully—you can identify the outline of the penis, meaning 2 must be the penile part of the urethra.
- Be careful when naming parts of the immature skeleton. Confusion arises between primary and secondary ossification centres. Current advice from the RCR (as of April 2019) recommends: *'Candidates must be able to identify all the different parts of the growing bone and you should be able to distinguish between epiphyses, apophyses and epiphyseal growth plates. Candidates who describe an epiphysis or apophysis as a secondary ossification centre, will lose one mark as this answer is only partially correct.'*

Q3.23 Coronal T1-weighted MRI of the pelvis

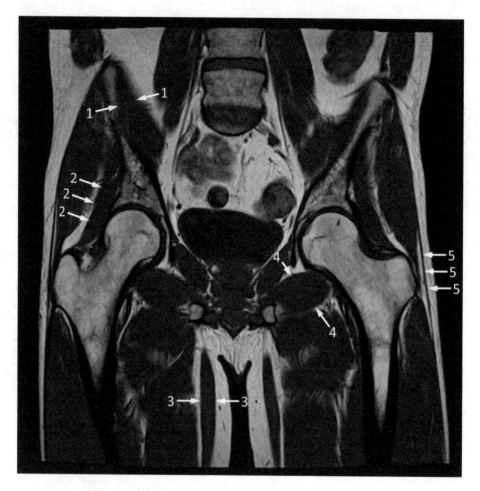

1. Name the arrowed structure.
2. Name the arrowed structure.
3. Name the arrowed structure.
4. Name the arrowed structure.
5. Name the arrowed structure.

Answers
1. Right iliacus muscle.
2. Right gluteus minimus muscle.
3. Right gracilis muscle.
4. Left obturator externus muscle.
5. Left fascia iliaca.

Comments:
This is a difficult question. The muscles of the pelvis are challenging and take a while to get your head around! One way the difficulty of the exam questions can be increased is by asking you to name common structures displayed in planes that you are not used to seeing. Knowing what structures lie medial or lateral can be useful, as can specific appearances of certain structures. A good example of this the feathery appearance of the **gluteus maximus** and **tensor fascia latae muscles**.

The **psoas major** and **iliacus muscles** combine to form the **iliopsoas muscle**, which descends along the medial iliac wing and eventually inserts into the **lesser trochanter on the femur**.

Exam tip:
- A common exam question is 'name a structure that attaches here', so knowing the insertions of common muscles is important. Having said that, some muscles have rather 'spread out' origins/insertions, which are difficult to definitively put an arrow on and these are perhaps less likely to come up in the exam. Examples of this include the origin and insertion of **pectineus** and the insertion of gluteus maximus.

Q3.24 Axial T1-weighted MRI of the pelvis

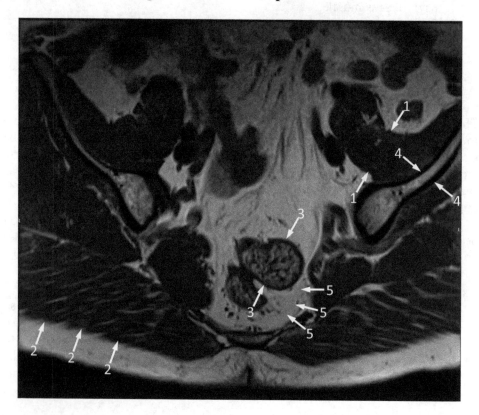

1. Name the arrowed structure.
2. Name the arrowed structure.
3. Name the arrowed structure.
4. Name the arrowed structure.
5. Name the arrowed structure.

Answers

1. Left iliopsoas muscle.
2. Right gluteus maximus muscle.
3. Rectum.
4. Left iliac bone.
5. Mesorectal fat.

Comments:

Remember that fat is bright on both conventional T1 and T2 sequences. Posterior and lateral to the iliac wings lie the three **gluteus muscles** (minimus, medius and maximus). The **gluteus maximus** has a feathery appearance, and which is the largest and most superficial of the three. The **mesorectal fat** surrounds the rectum and is an important structure when assessing local rectal cancer staging. The black 'dots' in the fat are predominantly small vessels.

Exam tip:

- Remember that flowing blood creates 'flow voids' on conventional spin-echo sequences. This is because the blood present in the vessels during slice selection has left the slice when it comes to read out and has been replaces by blood that was not 'excited'. Consequently, the blood in the vessels at read out returns no signal and appears black. This is useful in identifying vessels, for example the iliac vessels, which can be seen on this image relative to the greyer bowel loops lying more anteriorly. Note that if blood is slow-flowing or static, it does not leave the slice before the read out time and therefore returns signal and appears bright. Similarly, on sequences with short echo times (TE), such as gradient echo sequences, even fast flowing blood hasn't had a chance to leave the slice and still returns signal. This is why vessels often appear bright on gradient-echo sequences, whilst they appear dark on spin-echo sequences.

Q3.25 Axial T1-weighted MRI of the pelvis

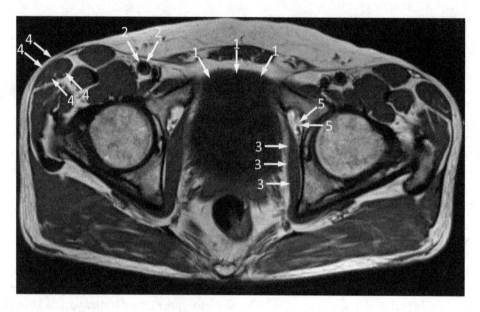

1. Name the arrowed structure.
2. Name the arrowed structure.
3. Name the arrowed structure.
4. Name the arrowed structure.
5. Name the arrowed structure.

Answers

1. Urinary bladder.
2. Right common femoral artery.
3. Left obturator internus muscle.
4. Right tensor fascia latae muscle.
5. Left obturator vessel.

Comments:

Obturator internus originates from the anterior-medial aspect of the pelvis and covers the majority of the **obturator foremen** (the hole surrounded by the **superior and inferior pubic rami**). It courses posteriorly along the inner surface of the pelvis to the **lesser sciatic foramen** at which point it is reflected by 90 degrees and continues towards the **greater trochanter**, where its tendon attaches.

The **obturator artery** (a terminal branch of the anterior trunk of the **internal iliac artery**) and **vein** are recognised by their position on the anterior pelvic side wall coursing towards the obturator foramen, which they pass through, and also by the presence of flow voids on this spin-echo sequence. Flow voids are better seen in the iliac vessels.

The **external iliac artery** (EIA) becomes the common femoral artery as it passes under the **inguinal ligament**, which runs from the **anterior superior iliac spine** to the **pubic tubercle**. On this image, the femoral heads are in view, so you can be confident you are caudal to the inguinal ligament and that this is the **common femoral artery** (CFA). The junction of the EIA and CFA can also be defined angiographically by the point of origin of the **inferior epigastric artery**, which courses superiorly along the anterior abdominal wall. Remember that the **tensor fascia latae** muscle can be recognised by its feathery appearance (similar to gluteus maximus). The **urinary bladder** is dark and fairly featureless on this T1 image. If this was a T2-weighted image, the wall would be roughly iso-intense to skeletal muscle and the urine would be very bright.

Exam tip:

- Remember, the order of the neuro-vascular bundle in the groin from lateral to medial is Nerve, Artery, Vein, (Y-fronts) [NAVY]. The artery also has a slightly thicker wall and is normally more spherical in cross section than the vein.

Q3.26 Coronal thick slice CT angiogram

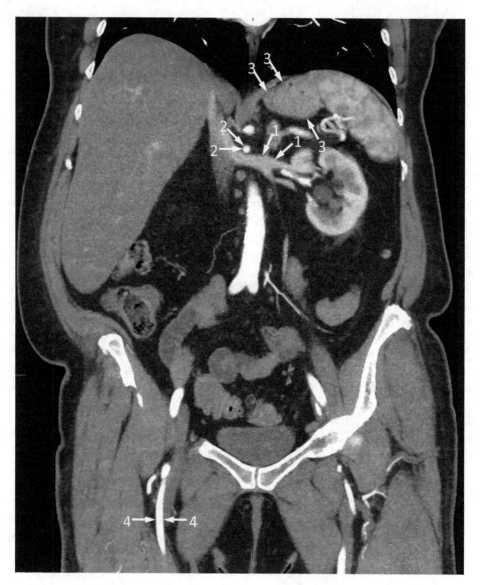

1. Name the arrowed structure.
2. Name the arrowed structure.
3. Name the arrowed structure.
4. Name the arrowed structure.
5. Name the anatomical variant.

Answers
1. Left renal vein.
2. Superior mesenteric artery.
3. Stomach.
4. Right superficial femoral artery.
5. Riedel lobe of the liver.

Comments:

The **left renal vein** passes just inferior to the origin of the **superior mesenteric artery** (SMA) to join the **inferior vena cava**. In a small number of individuals, the left renal vein passes posterior to the aorta, which is a normal anatomical variant known as a **retroaortic left renal vein**. Notice that the origin of the **coeliac axis** and SMA arise from the anterior margin of the aorta, which on coronal imaging gives them the appearance of a shotgun end-on. The SMA courses inferiorly very soon after its origin, meaning that the left renal vein (and 3rd part of the duodenum) is surrounded by the SMA superiorly and anteriorly, with the aorta posteriorly. If this passage is very tight it can cause compression of these structures resulting in 'nutcracker syndrome' with stenosis of the left renal vein and/or duodenum.

Notice the mottled appearance of the **spleen** on this arterial phase imaging (due to the difference in timing of enhancement of the white and red pulps). On this image, the **stomach** could easily be mistaken for the medial aspect of the spleen, but if you look closely, you will notice its more uniform enhancement pattern and the two small locules of gas within it.

A **Riedel lobe** is a normal variant, where the caudal part of the right hepatic lobe extends below the most inferior costal cartilage.

The **superficial femoral artery** (SFA) arises from the bifurcation of the **common femoral artery**, which also gives rise to the **profunda femoris** artery. The SFA is larger and courses medially within the thigh and does not give off any major branches. The profunda is more lateral and travels deeper in the thigh, giving off multiple branches. On this image it can be seen on the right lying just lateral to the proximal SFA.

Exam tip:
- Remember the importance of recognising that this is an angiogram (aorta opacifies, the IVC does not—and read the question title) so you do not accidentally label a vessel as a vein rather than an artery.

Q3.27 Axial fat suppressed PD-weighted MRI of a male pelvis

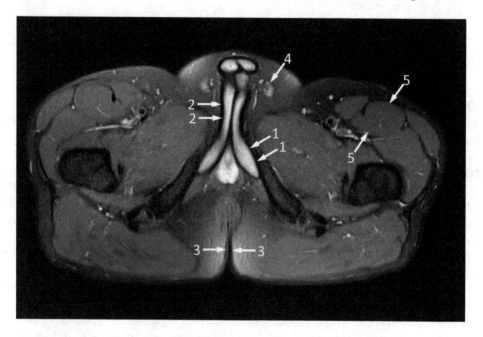

1. Name the arrowed structure.
2. Name the arrowed structure.
3. Name the arrowed structure.
4. Name the arrowed structure.
5. Name the arrowed structure.

Answers
1. Left penile crus.
2. Right corpus cavernosum.
3. Natal cleft (gluteal cleft; intergluteal cleft).
4. Left spermatic cord.
5. Left rectus femoris muscle.

Comments:

The bulk of the penis is made up of the paired **corpus cavernosa** and the **corpus spongiosum**. The corpus cavernosa originate at the pubic arch (well seen on this image) where they are known as the **penile crura** (paired). They course anteriorly and join together, where they continue as the corpus cavernosa. The corpus spongiosum originates at the **penile bulb** (seen here as the hyperintense structure between the two penile crura) and continues on the ventral aspect of the penis, encircling the penile urethra.

The **spermatic cord** contains the vas deferens, arteries, nerves, lymphatics, pampiniform plexus and tunica vaginalis.

The **rectus femoris** is seen at this level lying between **sartorius** (medial), **vastus lateralis** (posterolateral) and **tensor fascia latae** (anterolateral). It makes up one of the quadriceps group of muscles (along with vastus lateralis, intermedius and medialis), which all share a common tendon (**quadriceps tendon**).

Exam tip:
- Remember that this is a fat suppressed image, so anything that is bright is not fat! In this example, the penile corpora appear bright because they contain slow flowing blood. Similarly, the spermatic cords are fairly hyper-intense due to the presence of slow flowing blood within the pampiniform plexus, whilst the surrounding fat has been suppressed out.

Q3.28 Sagittal T2-weighted MRI of a female pelvis

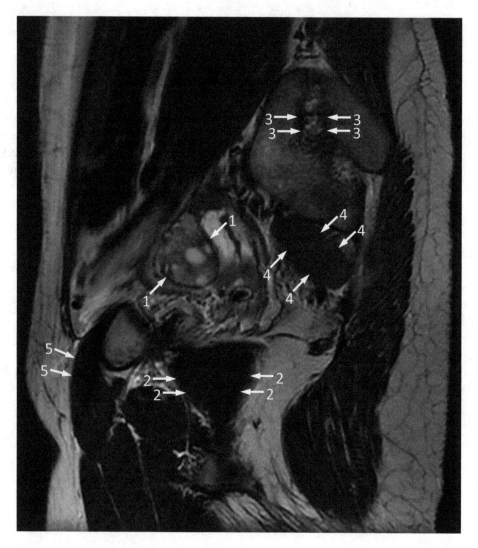

1. Name the arrowed structure.
2. Name the arrowed structure.
3. Name the arrowed structure.
4. Name the arrowed structure.
5. Name the arrowed structure.

Answers
1. Ovary.
2. Obturator internus muscle.
3. Sacro-iliac joint.
4. Piriformis muscle.
5. Pectineus muscle.

Comments:
The **ovary** is recognised on T2 images by the presence of hyper-intense (fluid filled) follicles, bound by a hypo-intense stroma.

Between the **superior** and **inferior pubic rami** lies the **obturator foramen**, which is covered by a membrane. The **obturator internus** and **externus** lie on the inner and outer surface of this foramen, as demonstrated here. Note the **obturator vessels** traversing the foramen at its antero-superior margin.

Piriformis is a pyramid-shaped muscle originating from the anterior aspect of the sacrum and **greater sciatic notch**. It exits the pelvis through the **greater sciatic foramen** (the space between the greater sciatic notch and the **sacrospinous ligament**) and inserts onto the **greater trochanter**. **Pectineus** is a fairly flat, quadrangular shaped muscle and is the most anterior of the hip adductors (although its main action is hip flexion). It originates from the pectineal line on the anterior margin of the pubic bone and courses laterally and inferiorly to attach at the medial boarder of the femur just inferior to the **lesser trochanter**.

Exam tip:
- This question is an example of how structures that you would easily recognise on an axial image can appear very different and difficult to identify in another plane. The best way to start is to orientate yourself by recognising bony landmarks. In this example the pubic rami and SI joints are useful in recognising this a para-median image. Remember, there is no way to work out whether this image is left or right, so don't go there!

Q3.29 Coronal CT angiogram

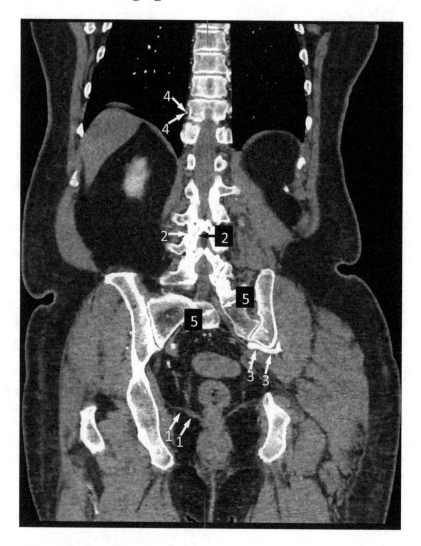

1. Name the arrowed structure.
2. Name the arrowed structure.
3. Name the arrowed structure.
4. Name the arrowed structure.
5. Name the arrowed structure.

Answers
1. Right levator ani muscle.
2. Right L4 pars interarticularis.
3. Left superior gluteal artery.
4. Right T12 intercostal artery.
5. Left S2 neural foramen.

Comments:
The **levator ani** muscle is a broad, slender muscular sling made up of 3 constituent parts (pubococcygeus, puborectalis and iliococcygeus) although conclusively delineating these on imaging is difficult. It originates from the inner wall on the inferior pelvic margin and extends medially and inferiorly to blend with the rectum and external anal sphincter. On coronal images it often has a 'seagull' like appearance.

The **pars interarticularis** is the small part of bone between the superior and inferior articular processes. It is important to know about as 'pars defects' are a common cause of anterolisthesis of L5 on S1 in a degenerative spine. Note that identifying vertebral levels here might seem difficult, but if you recognise the superior articular process of S1 on the right, you can count up from there, identifying question 2 as L4 and question 4 as T12.

The **sacral neural foramina** course much more vertically than at the lumbar levels.

The **superior gluteal artery** is the largest branch of the posterior trunk of the **internal iliac artery** (IIA). It exits the pelvis through the **greater sciatic foramen**, just superior to **piriformis** (which can be seen on this image). The **inferior gluteal** and **internal pudendal** arteries (both branches of the anterior trunk of the IIA), also exit the pelvis through the greater sciatic foramen, but lie inferior to piriformis.

Exam tip:
- Remember to read the title of the image, which here states this is an angiogram. This allows you to recognise that the hyperdense paired circular structures passing just inferior to the **costovertebral joints** on this image must be **intercostal arteries**.

Q3.30 Sagittal T2-weighted MRI of a female pelvis

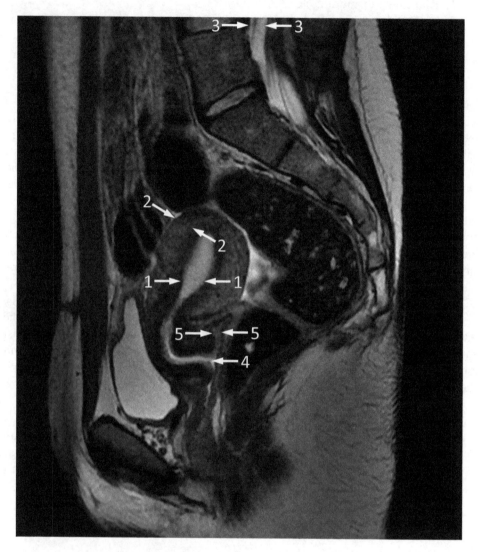

1. Name the arrowed structure.
2. Name the arrowed structure.
3. Name the arrowed structure.
4. Name the arrowed structure.
5. Name the arrowed structure.

Answers
1. Endometrium.
2. Outer myometrium.
3. Thecal sac.
4. External cervical os.
5. Posterior vaginal fornix.

Comments:
The uterine anatomy is very well demonstrated on T2-weighted MRI. The inner-most hyperintense layer is the **endometrium**. There is then a hypointense band between the bright endometrium and the slightly less bright, heterogenous **outer myometrium**, known as the **junctional zone.**

The inferior part of the **cervix** protrudes into the upper vaginal vault, forming the anterior and posterior **vaginal fornices**. The **endometrial canal** is in communication with the vaginal vault via the **endocervical canal**, with the uterine and vaginal orifices of this canal known as the **internal** and **external cervical os,** respectively. Note that the uterus in this image is retroflexed.

Exam tip:
- Uterine zonal anatomy is recognised on sagittal images, but it is important to familiarise yourself with it on axial and coronal planes too. Remember that this is a T2-weighted image meaning that fluid, such as CSF within the **thecal sac**, is bright.

Q3.31 Coronal T2-weighted MRI of a female abdomen

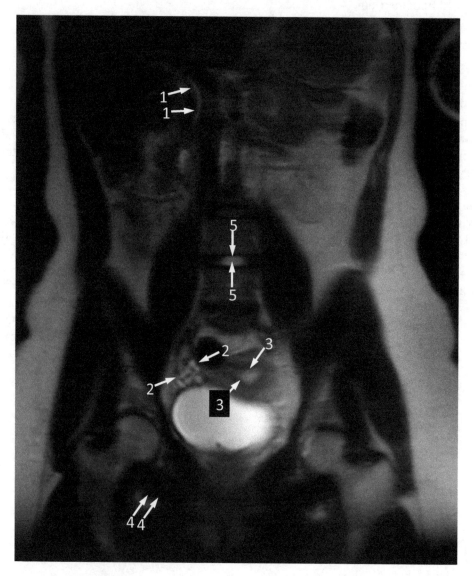

1. Name the arrowed structure.
2. Name the arrowed structure.
3. Name the arrowed structure.
4. Name the arrowed structure.
5. Name the arrowed structure.

Answers
1. Common bile duct.
2. Right ovary.
3. Endometrium.
4. Right obturator externus muscle.
5. Nucleus pulposus of a lumbar intervertebral disc.

Comments:
The **common bile duct** (CBD) is formed by the confluence of the **common hepatic duct** (itself formed from the left and right hepatic ducts) and **cystic duct**. It typically lies just inferior to the portal vein before entering the pancreas, where in it joins with the **main pancreatic duct** before draining into the 2nd part of duodenum at the **Ampulla of Vater**.

As discussed in question 8, the **ovaries** are readily identified on T2-weighted images by their multiple hyper-intense follicles. The junctional anatomy of the uterus discussed in question 10 is again demonstrated here, but this time in a coronal plane. **Obturator externus** can be clearly seen from its origin at the obturator foramen coursing towards its insertion at the intertrochanteric fossa on the proximal femur.

The **intervertebral discs** consist of a central **nucleus pulposus**, which is bound by an outer **annulus fibrosus**. The nucleus pulposus is hydrated and so bright on T2 images, whilst the annulus is fibrous and hence dark. In degeneration, the discs dehydrate and the nucleus becomes increasingly iso-intense.

Exam tip:
- Remembering that fluid is bright on T2-weighted sequences is essential in identifying the CBD. This is a relatively low spatial resolution, large field of view image, quite typical of a standard abdominal MR protocol. Identifying the iliac crests initially appears challenging, however, if you remember the bony medulla is relatively fatty and hence bright on this non-fat saturated image, you will be able to identify the iliac crest, particularly on the right. Once you have identified this, identifying the inferior pubic rami is relatively straight forwards and, by inference, the location of the obturator foramen. As discussed in question 8, you should then be able to identify obturator internus and externus.

Q3.32 Axial CT of a male pelvis

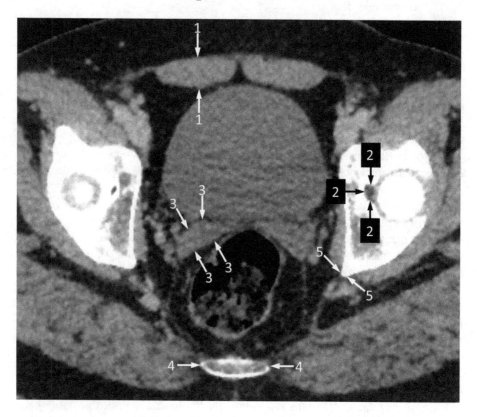

1. Name the arrowed structure.
2. Name the arrowed structure.
3. Name the arrowed structure.
4. Name the arrowed structure.
5. Name the arrowed structure.

Answers
1. Right rectus abdominus muscle.
2. Left acetabular fossa.
3. Right seminal vesicle.
4. Coccyx.
5. Left ischial spine.

Comments:
The paired **rectus abdominus** muscles are readily recognisable on axial images. Their fibrous capsules blend with the **linea alba** aponeurosis medially and the **linea semilunaris** laterally.

The **acetabulum** is formed by the confluence of the **ilium, ischium** and **pubis.** In the developing skeleton, the ossification centre at the point where these three bones meet is called the **triradiate cartilage.** At the centre of the mature acetabulum lies a depression called the **acetabular fossa.** This contains the **ligament of the femoral head (ligamentum teres femoris),** which attaches the **fovea of the femoral head** to the acetabulum. It also contains fat, which explains its dark appearance on CT.

The **seminal vesicles** lie infero-posterior to the urinary bladder just superior to the prostate and have a 'bow tie' appearance on axial images.

The **sacrospinous ligament** attaches to the **ischial spine** laterally and the lateral border of the **sacrum** and **coccyx** medially. It acts to stabilise the ilium and sacrum from rotating about each other. It's fibres also divide the **greater sciatic foramen** superiorly from the **lesser sciatic foramen** inferiorly.

Exam tip:
- Fat is very useful when defining structures on CT. Fat planes, such as those seen here between the bladder and seminal vesicles are useful to define the borders of a structure. If you are having difficulty working out what a structure is, have a look at the fat planes around it to see if that helps to separate it from other neighbouring structures.

Q3.33 Coronal contrast-enhanced, fat supressed T1-weighted MRI

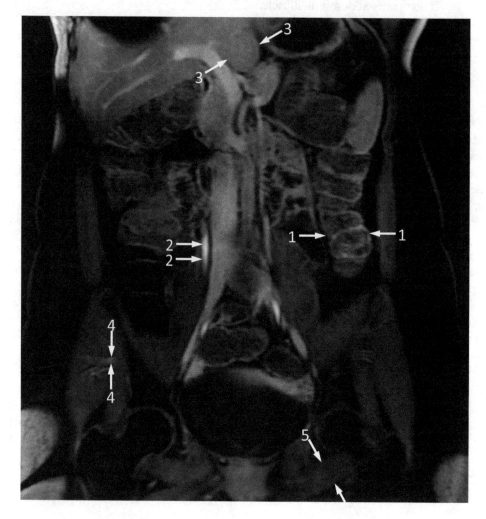

1. Name the arrowed structure.
2. Name the arrowed structure.
3. Name the arrowed structure.
4. Name the arrowed structure.
5. Name the arrowed structure.

Answers
1. Descending colon.
2. Right ureter.
3. Caudate lobe of the liver.
4. Right superior gluteal artery.
5. Left obturator externus muscle.

Comments:

The colon is identified both by its location, but also by the appearance of faecal material within the lumen. Compare this to the more centrally located loops of small bowel seen on this image.

After leaving the renal pelvis, the **ureters** lie just anterior to the **psoas major** muscles in a para-midline location until they enter the pelvis, at which point their course is more tortuous until they enter the **urinary bladder** at the vesicoureteral junction.

The **superior gluteal artery** is a branch of the posterior division of the **internal iliac artery**. It exits the pelvis through the **greater sciatic foramen** above **piriformis** and lies between **gluteus minimus** and **medius** (as seen in this image).

Exam tip:
- The ureters are seen as bright structures on this image due to the phase of the scan (relatively late portal venous phase) which means that some contrast has already been excreted by the kidneys. It would be easy to mistake the right ureter for a vessel, however, there are no major vascular branches in this region that could be readily identified on a single image.

Q3.34 Selective mesenteric angiogram

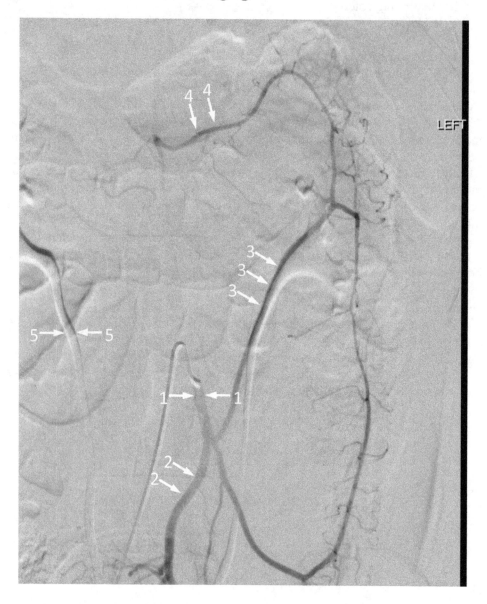

1. Name the arrowed structure.
2. Name the arrowed structure.
3. Name the arrowed structure.
4. Name the arrowed structure.
5. Name the arrowed structure.

Answers
1. Inferior mesenteric artery.
2. Superior rectal artery.
3. Left colic artery.
4. Marginal artery of Drummond.
5. Right ureter.

Comments:
The **inferior mesenteric artery** (IMA) is the 3rd major anterior branch of the abdominal aorta, although it often arises slightly off the midline. It supplies the colon from the **splenic flexure** to the superior **rectum**. It has three major branches, the **superior rectal artery**, sigmoid branches and the **left colic artery**. The left colic artery usually continues as the **marginal artery of Drummond** to anastomose with branches of the **middle colic artery**, a branch of the **superior mesenteric artery**.

Exam tip:
- Recognising this as a selective angiogram of the IMA is essential. There are several clues on the image to help you: The side is labelled and, if you look closely, you can see ghosting of the descending and transverse colon, as well as both renal collecting systems. This should help you to identify that the vessel selected with the angiographic catheter lies relatively low in the abdomen (at approximately L3). You know from the question title this is a *mesenteric* angiogram and the origin of this vessel is too caudal to be the coeliac axis or SMA, therefore it has to be the IMA.

Q3.35 Axial T2-weighted MRI

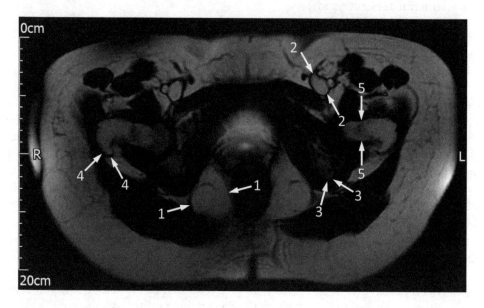

1. Name the arrowed structure.
2. Name the arrowed structure.
3. Name the arrowed structure.
4. Name the arrowed structure.
5. Name the arrowed structure.

Answers
1. Right ischioanal fossa.
2. Left common femoral vein.
3. Left ischial tuberosity.
4. Greater trochanter of the right femur.
5. Neck of the left femur.

Comments:

The **ischioanal fossa** is a fat-filled wedge-shaped cavity lying lateral to the anal canal and inferior to the urogenital diaphragm.

The **external iliac** vessels become the **common femoral** vessels as they pass under the **inguinal ligament**. This can be difficult to determine radiologically and so the origin of the **inferior epigastric** vessels is a useful landmark to define the junction. In this image, the femoral necks are seen so you can be confident the level is inferior to the inguinal ligament. Question 2 must therefore be a common femoral vessel. The relationship of the external iliac/common femoral neurovascular bundle, from lateral to medial, is Nerve, Artery, Vein (remembered by the pneumonic NAVy). Another clue is the smaller calibre of the artery, which also appears more circular in cross-section; the vein is typically larger and more elliptical by comparison.

Exam tip:
- This is an example of a question showing structures labelled on an MR sequence that you may not be familiar with. In the exam, common structures may be labelled on sequences you are less familiar with, or in unusual planes (e.g. identifying obturator internus on a coronal image). The image in this question is a balanced steady-state gradient echo sequence, which is essentially a heavily T2-weighted gradient echo. It may be referred to by manufacturer proprietary names such as FIESTA or CISS, and is more commonly encountered in neuroradiological images to assess the CSF spaces. The vessels do not have a typical 'flow void' within them because this is a gradient echo rather than a spin echo sequence. Whatever the sequence or modality, the anatomy is the same and taking some time to identify a 'landmark', such as the femoral head in this case, and working out the other structures from there is a helpful technique.

Q3.36 Coronal paediatric dual phase contrast-enhanced CT

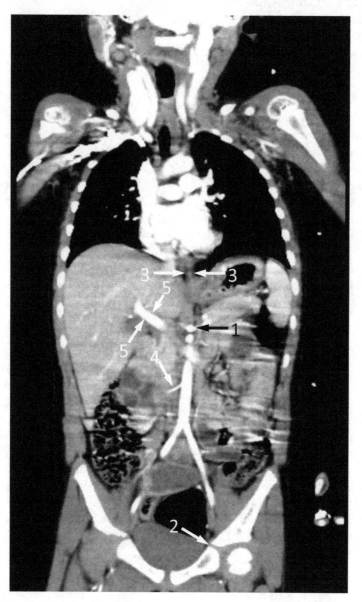

1. Name the arrowed structure.
2. Name the arrowed structure.
3. Name the arrowed structure.
4. Name the arrowed structure.
5. Name the arrowed structure.

Answers
1. Coeliac axis.
2. Left triradiate cartilage.
3. Oesophagus.
4. Right renal artery.
5. Portal vein.

Comments:
The **coeliac axis** is the most superior anterior branch of the abdominal aorta. The coeliac axis and **superior mesenteric artery** have a characteristic 'shotgun end on' appearance when viewed in a coronal plane. In a sagittal plane, you can recognise the coeliac axis by its more superior position, but also by its more horizontal course, whilst the superior mesenteric artery courses inferiorly soon after its origin. The 3rd part of the duodenum and left renal vein pass between the aorta and SMA.

The **triradiate cartilage** is an important structure to recognise in paediatric patients. It is a non-articulating epiphyseal plate where the ischium, ilium and pubis meet in the developing skeleton. It typically fuses in mid-adolescence and should not be mistaken for a fracture.

Exam tip:
- The **portal vein** is readily identified on coronal imaging. It is important to note the phase of contrast of this CT—it has been performed as a dual bolus CT (modified Bastion protocol), which involves an initial bolus of contrast, followed by a delay of around 30 s, and then a second bolus just before image acquisition. This gives a combined portal venous and arterial phase and is commonly used for trauma CT. This is why both the portal vein, as well as the arterial tree, are well opacified with contrast.

Q3.37 Paediatric transverse ultrasound of the upper abdomen

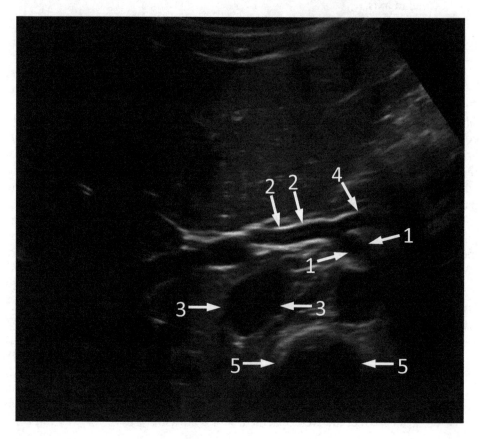

1. Name the arrowed structure.
2. Name the arrowed structure.
3. Name the arrowed structure.
4. Name the arrowed structure.
5. Name the arrowed structure.

Answers
1. Coeliac axis.
2. Common hepatic artery.
3. Inferior vena cava.
4. Splenic artery.
5. Vertebral body.

Comments:
On axial (or transverse) imaging, the **coeliac axis** classically has a seagull or 'T-shaped' appearance, as seen here. It typically lies at the level of T12 and branches initially into the **left gastric, splenic** and **common hepatic arteries**. The common hepatic artery gives off the **gastroduodenal artery** (which is an anastomotic link to the superior mesenteric artery) and then the **right gastric artery**. After this point, the vessel is known as the **proper hepatic artery** and continues into the hepatic hilum to supply the liver.

Exam tip:
- Ultrasound images can initially seem hard to interpret. However, it is important to remember that any structure labelled in the exam must be readily identifiable on the single image provided. This limits the structures that could potentially be examined on an USS image to those that have a typical distinctive appearance, such as the coeliac axis here. Another example may be a transverse image of the neck at the level of the thyroid of a longitudinal section through a neonatal hip.

IOP Publishing

Anatomy for the Royal College of Radiologists Fellowship

Illustrated questions and answers

**Andrew G Murchison, Mitchell Chen, Thomas Frederick Barge, Shyamal Saujani,
Christopher Sparks, Radoslaw Adam Rippel, Malcolm Sperrin and Ian Francis**

Chapter 4

Musculoskeletal system

Shyamal Saujani and Radoslaw Adam Rippel

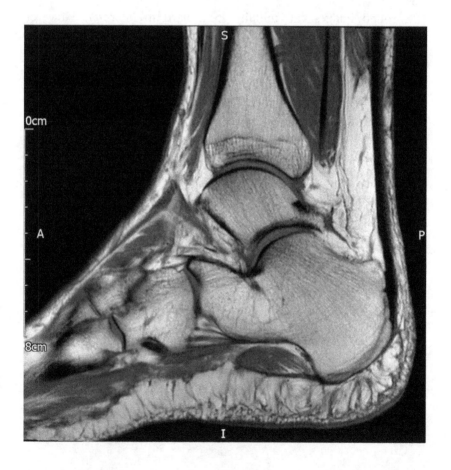

Q4.1 Plain radiograph of a foot in AP projection

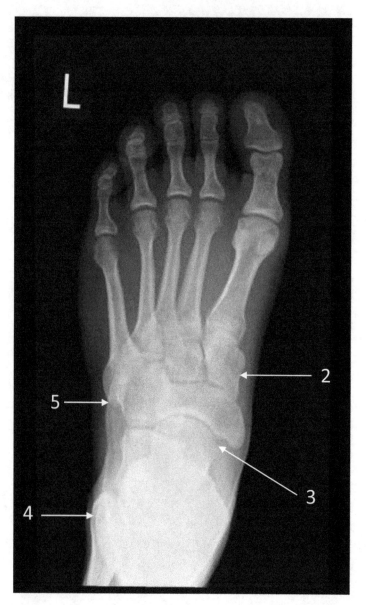

1. Name the anatomical variant.
2. Name the arrowed structure.
3. Name the arrowed structure.
4. Name the arrowed structure.
5. Name the structure which attaches to this structure.

Answers
1. Os naviculare (accessory navicular).
2. Left medial cuneiform bone.
3. Head of the left talus.
4. Left lateral malleolus.
5. Left peroneus brevis tendon.

Comments:

Anatomical variants are an important part of the anatomy exam as well as day-to-day clinical practice. An ability to distinguish between an acute fracture and normal variant may have significant impact on patient care and clinical outcome.

Anatomical variants may be subtle at times. The question image shows a relatively subtle **Os naviculare**. Below is a more prominent example. Can you name any more ossicles of the foot?

Bones of the forefoot include the **medial, intermediate** and **lateral cuneiforms, cuboid, navicular, talus** and **calcaneus.** Knowing the relationship between the bones will help you to orientate yourself. The **navicular** and **medial cuneiform** bones are on the medial aspect of the foot whilst the **cuboid** is on the lateral side.

Exam tip:
- Question 5 specifically asks about a structure attaching to the **5th metatarsal** at the point indicated. Knowledge of tendon attachments is useful for picking up subtle avulsion fractures as well as reporting soft tissue injuries seen on MR imaging.

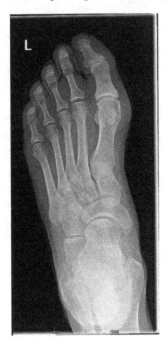

Q4.2 Axial T2-weighted section from a MRI of the lumbar spine

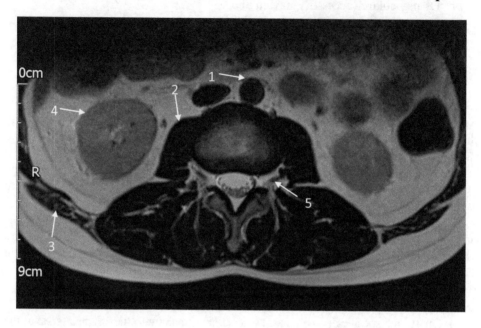

1. Name the arrowed structure.
2. Name the arrowed structure.
3. Name the arrowed structure.
4. Name the arrowed structure.
5. Name the arrowed structure.

Answers
1. Aorta.
2. Right psoas major muscle.
3. Right latissimus dorsi muscle.
4. Right kidney.
5. Left exiting nerve root.

Comments:
Although the image is from MRI spine series, there are a number of pertinent strucutres around the **spinal column**. All of these should be reviewed while reporting on spinal imaging.

Each lumbar vertebra is composed of a **vertebral body,** paired **pedicles, lamina, facet joints** and **transverse process** and a **spinous process**. The spinous process can be bifid, which is a normal anatomical variant.

Exam tip:
- It can be helpful to practice looking at standard anatomical structures in unusual views and modalities. Although this is an MRI of the lumbar spine, a number of important intra-abdominal sturctures can be identified on the image.

Q4.3 Coronal T1-weighted section from an MRI of the knee

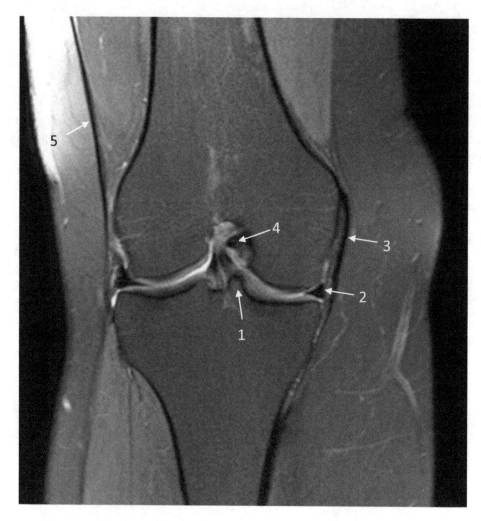

1. Name a structure which attaches to this structure.
2. Name the arrowed structure.
3. Name the arrowed structure.
4. Name the arrowed structure.
5. Name the arrowed structure.

Answers
1. Anterior cruciate ligament.
2. Medial meniscus.
3. Medial collateral ligament.
4. Posterior cruciate ligament.
5. Iliotibial band.

Comments:
Familiarise yourself with soft tissue structures around the knee joints. Are you able to identify them in **axial** and **sagittal** views?

The **anterior cruciate ligament** inserts into tibia anteriorly and medially. The **posterior cruciate ligament** inserts onto the postero-lateral aspect of the tibia. Note how thick and round the PCL is in comparison to more sheet-like ACL, which is best appreciated on sagittal views.

The **medial** and **lateral menisci** are C-shaped cushions positioned between the **femoral condyles** and medial and lateral **tibial plateaus**. On sagittal images, the have a 'bow-tie' appearance with **anterior** and **posterior horns**.

Exam tips:
- The iliotibial band is a fascial band that is continuous with the fascia of the tensor fasciae latae muscle. It is on the lateral aspect of the thigh, and can be used to distinguish the lateral and medial sides of the knee on coronal images.
- Make sure you can identify all tendons and ligaments passing through and inserting around the knee joint. You may be asked to identify a structure in any view—axial, coronal, sagittal.

Q4.4 Coronal section from a CT of the abdomen and pelvis

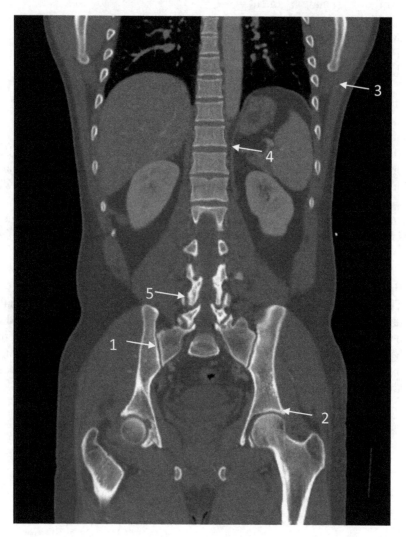

1. Name the arrowed structure.
2. Name the arrowed structure.
3. Name the arrowed structure.
4. Name the arrowed structure.
5. Name the arrowed structure.

Answers
1. Right sacro-iliac joint.
2. Anterior rim of the left acetabulum.
3. Left latissimus dorsi muscle.
4. Left crus of the diaphragm.
5. Right L4/5 facet joint P.

Comments:
Each hemipelvis comprises of three fused bones: the **ilium, ischium** and **pubis**. The **acetabulum** is a cup-like depression formed by all three bones which articulates with the **head of the femur**. The **labrum** of the acetabulum is a fibrous structure which deepens the acetabulum—it is not well demonstrated on CT but can be appreciated on MRI.

The pelvis articulates posteriorly with the **sacrum** at the **sacroiliac joints**. The two sides of the pelvis meet anteriorly at the **pubic symphysis.**

Exam tip:
- Facet joints are important anatomical landmark and target for steroid injections. Can you identify them on a plain film? In addition make sure you can point out **transverse process**, and **spinous process**.

Q4.5 Cervical spine radiograph, lateral view

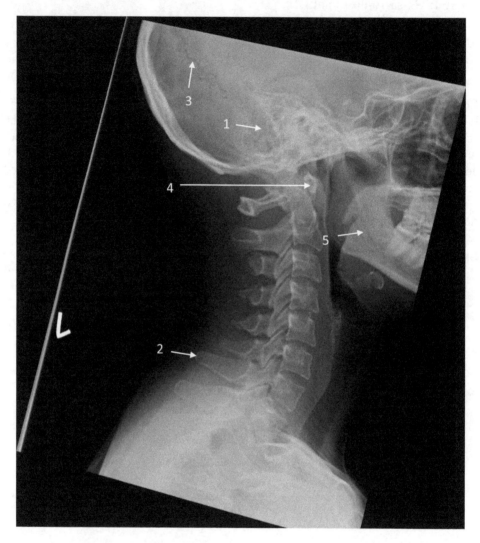

1. Name the arrowed structure.
2. Name the arrowed structure.
3. Name the arrowed structure.
4. Name the arrowed structure.
5. Name the arrowed structure.

Answers
1. Mastoid air cells.
2. Spinous process of C7.
3. Lambdoid suture.
4. Anterior arch of C1.
5. Mandibular canal.

Comments:
The C1 and C2 vertebra are different in structure from the other cervical vertebrae.

  C1 (the atlas) is a ring-like structure. It has an **anterior arch** and a **posterior arch**, with two **lateral masses** between them. The lateral masses contain the articular surfaces for the **occipital condyles** superiorly, forming the **atlanto-occipital joints**. The articular surfaces where C1 meets C2 at the **atlanto-axial joints** are positioned on the inferior border of the lateral masses. C1 also has two **transverse processes** lateral to the lateral masses, with **transverse foramina** allowing passage of the **vertebral arteries**.

  The **odontoid process (peg)** arises from the body of **C2 (the axis)**. The **transverse ligament** passes posterior to the odontoid process to secure its articulation with the anterior arch of C1.

Exam tip:
- When naming one of the cervical vertebrae, it can be helpful to start with C2 and count down.

Q4.6 AP radiograph of the knee

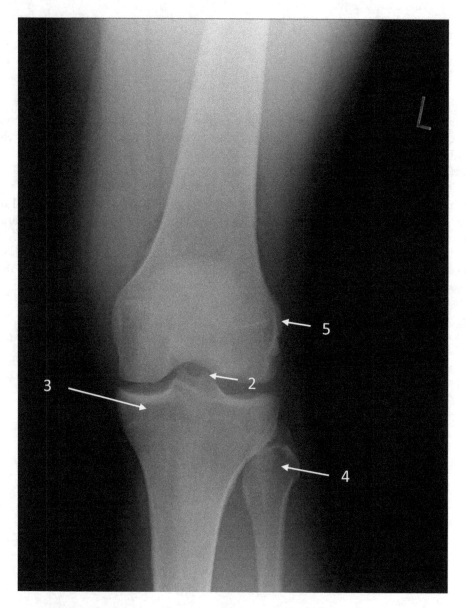

1. Name the anatomical variant.
2. Name a structure which attaches to this structure.
3. Name the arrowed structure.
4. Name the arrowed structure.
5. Name the arrowed structure.

Answers
1. Fabella.
2. Anterior cruciate ligament.
3. Medial tibial plateau.
4. Head of the fibula.
5. Lateral femoral condyle.

Comments:

Anatomical variants may be subtle at times, as with the **fabella** shown here. Below is a lateral radiograph of the same patient.

A **bipartite patella** is another common anatomical variant in the knee, which forms from an accessory ossification centre. There are also variants in appearance of the meniscus such as **discoid meniscus**.

Exam tip:
- Question 2 specifically asks about a structure attaching at the medial tibial spine. The most common question format in the exam will ask you to name the arrowed structure. However, different formats will also crop up, so it's important to read each question before answering.

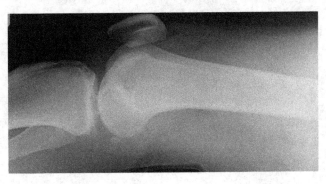

Q4.7 Axial fat-suppressed STIR sequence from an MRI of the knee

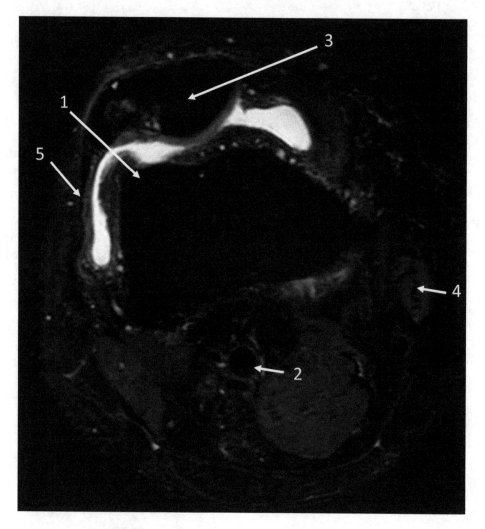

1. Name the arrowed structure.
2. Name the arrowed structure.
3. Name the arrowed structure.
4. Name the arrowed structure.
5. Name the arrowed structure.

Answers

1. Lateral femoral condyle.
2. Popliteal artery.
3. Patella.
4. Sartorius muscle.
5. Lateral patellofemoral retinaculum.

Comments:

Fat signal is suppressed on this STIR sequence, meaning that bones and subcutaneous fat appear dark.

The **lateral patellofemoral retinaculum** is a fibrous structure between the **iliotibial band** and the **patella**.

The **sartorius** is one of three muscles whose tendons join to form the pes anserinus, which inserts onto the medial side of the tibia—the other two are the gracilis and semitendinosus muscles.

The question image actually demonstrates a **bipartite patella,** but this is very difficult to spot on MRI. Below is a more prominent example on a plain radiograph.

Exam tip:

- The lateral femoral condyle has a higher 'peak' than the medial femoral condyle on axial images. This can help orientate you to which side is medial and which is lateral.

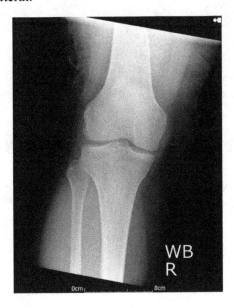

Q4.8 Sagittal T2-weighted MRI of the ankle

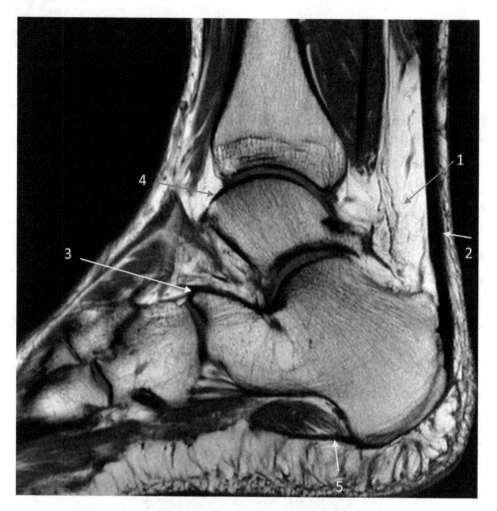

1. Name the arrowed structure.
2. Name the arrowed structure.
3. Name the arrowed structure.
4. Name the arrowed structure.
5. Name the arrowed structure.

Answers
1. Kager fat pad (precalcaneal or preachilles fat pad).
2. Achilles tendon.
3. Anterior process of the calcaneus.
4. Dome of the talus.
5. Plantar fascia.

Comments:
Ankle anatomy is complex with multiple muscles, tendons and vasculature passing anteriorly and posteriorly to the bony structures.

The tibiotalar joint is the articulation of the **dome of the talus** (the articular surface of the **body of the talus**) with the **tibia** (including the **medial malleolus**) and the **lateral malleolus of the fibula**. The anterior part of the talus is called the **head**, and this articulates with the **navicular bone**. The head and body of the talus are joined by the **neck of the talus**.

The **Achilles (calcaneal) tendon** is formed from the **soleus** and **gastrocnemius muscles**, and attaches to the **calcaneal tuberosity**. The calcaneus articulates with the **cuboid** anteriorly and the talus superiorly—the **sustenaculum tali** is a bony process arising from the medial calcaneus which forms part of the **talocalcaneal joint**. A cavity known as the **sinus tarsi** (tarsal sinus) is found between the two bones on the lateral aspect of the foot.

Exam tip:
- Tendons appear dark on T1 and T2 sequences. Remember to state whether the arrow is pointing to a muscle or a tendon for full marks.

Q4.9 AP plain radiograph of the pelvis

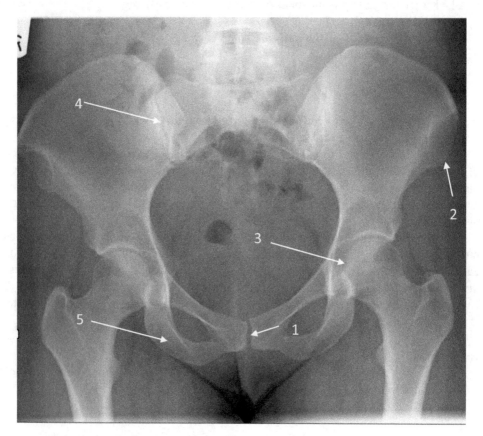

1. Name the arrowed structure.
2. Name one structure that inserts here.
3. Name the arrowed structure.
4. Name the arrowed structure.
5. Name the arrowed structure.

Answers
  1. Symphysis pubis (pubic symphysis).
  2. Left sartorius muscle or left inguinal ligament (the arrow points to the anterior superior iliac spine).
  3. Left fovea capitis.
  4. Right sacro-iliac joint.
  5. Right inferior pubic ramus.

Comments:

The **obturator foramen** on each side of the pelvis allow passage of the obturator artery, vein and nerve. The **superior pubic ramus** lies above the obturator foramen and the **inferior pubic ramus** below it.

Exam tip:
  • The sites at which muscles attach to the pelvic bones are useful to learn as they are readily testable.

Q4.10 Coronal T1-weighted sequence from an MRI of the knee

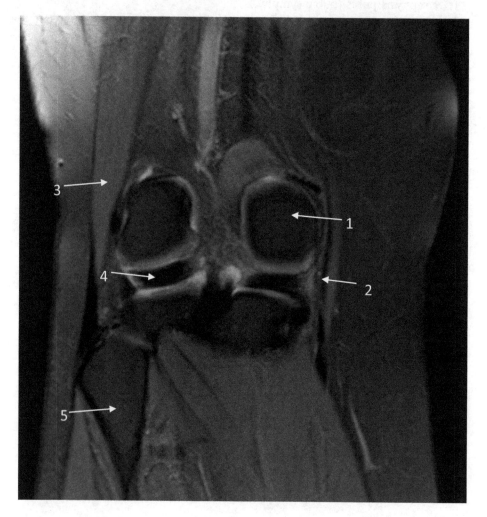

1. Name the arrowed structure.
2. Name the arrowed structure.
3. Name the arrowed structure.
4. Name the arrowed structure.
5. Name the arrowed structure.

Answers
1. Medial femoral condyle.
2. Medial collateral ligament.
3. Biceps femoris muscle.
4. Lateral meniscus.
5. Neck of fibula.

Comments:

The **biceps femoris muscle** is one of the hamstring group in the posterior compartment of the thigh. It has a **long head** which attaches to the **ischial tuberosity**, and a short head which attaches to the shaft of the femur. Its tendon joins with the **lateral collateral ligament** to form the conjoint tendon which attaches to the **head of the fibula**.

The medial hamstring muscles, the **semimembranosus** and the **semitendinosus**, also attach to the ischial tuberosity. The semimembranosus muscle inserts onto the **medial femoral condyle**, and the tendon of semitendinosus forms the pes ansirinus with the **sartorius** and **gracilis** tendons, and inserts onto the medial **tibia**.

Exam tips:
- This is another example of common and familiar structures seen in a slightly more challenging projection. Review knee MRIs in axial, sagittal and coronal views to understand how different structures relate to each other.
- The fibula and the iliotibial band are both lateral structures that can help to orientate you within the image.

Q4.11 Lateral radiograph of the ankle

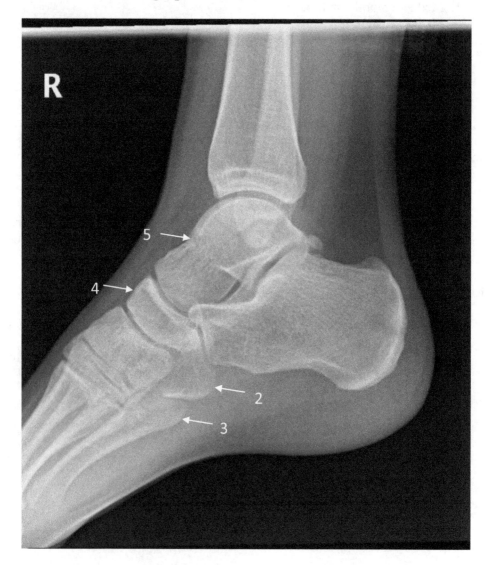

1. Name the anatomical variant.
2. Name the arrowed structure.
3. Name the arrowed structure.
4. Name the arrowed structure.
5. Name the arrowed structure.

Answers
1. Os trigonum.
2. Cuboid bone.
3. Base of the 5th metatarsal.
4. Navicular bone.
5. Neck of the talus.

Comments:
The **base of the 5th metatarsal** is an important review area where subtle fractures are often missed. This is particularly the case when looking at plain radiographs of the ankle where the base of the 5th metatarsal is seen on the edge of the film.

Exam tip:
- Foot ossicles are a group of anatomical variants that lend themselves very well to exam questions, and are well worth learning.

Q4.12 Axial T2-weighted section from an MRI of the foot

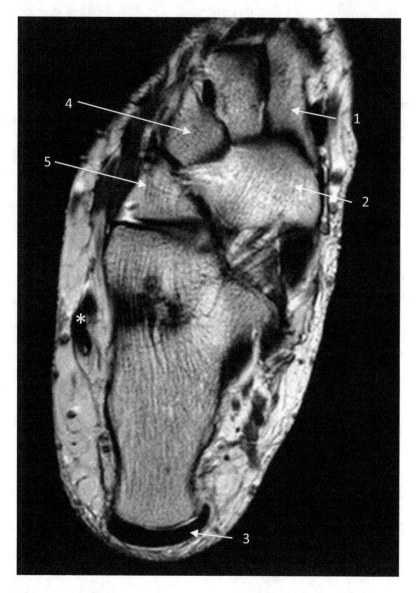

1. Name the arrowed structure.
2. Name the arrowed structure.
3. Name the arrowed structure.
4. Name the arrowed structure.
5. Name the arrowed structure.

Answers
1. Medial cuneiform bone.
2. Navicular bone.
3. Achilles tendon.
4. Lateral cuneiform bone.
5. Cuboid bone.

Comments:
The small bones of the foot may be confusing and difficult to identify. Try to start by identifying the lateral and medial aspect of the foot—you can look out for characteristic alignment of **cuneiform bones**, or the unique shape of **the navicular** or **cuboid bones.**

Exam tip:
- Distinguishing medial from lateral is important in images of the axial skeleton. The peroneus tendons (marked with a yellow asterisk in the image for this question) pass behind the lateral malleolus and continue along the lateral side of the calcaneus—they can therefore be useful in confirming the lateral side of the image.

Q4.13 Sagittal fat-suppressed T1-weighted section from an MRI of the knee

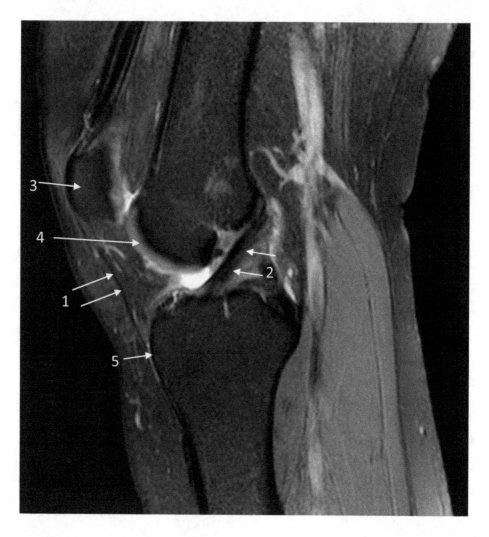

1. Name the arrowed structure.
2. Name the arrowed structure.
3. Name the arrowed structure.
4. Name the arrowed structure.
5. Name the arrowed structure.

Answers
1. Patellar tendon.
2. Anterior cruciate ligament.
3. Patella.
4. Cartilage over femoral condyle.
5. Tibial tuberosity.

Comments:
The tendons of the quadriceps muscles in the anterior thigh combine to form the **quadriceps tendon,** which inserts onto the superior border of the **patella**. The **patellar tendon** arises from the lower border of the patella and inserts onto the **tibial tuberosity**.

Exam tip:
- Remember that the anterior cruciate ligament inserts anteriorly on the tibia, and the posterior cruciate ligament inserts posteriorly.

Q4.14 Oblique plain radiograph of the foot

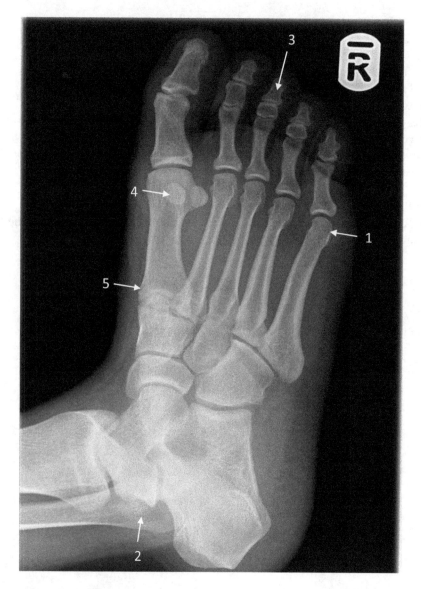

1. Name the arrowed structure.
2. Name the arrowed structure.
3. Name the arrowed structure.
4. Name the arrowed structure.
5. Name the arrowed structure.

Answers
1. Head of the 5th metatarsal.
2. Lateral malleolus.
3. Distal phalanx of the 3rd toe.
4. Sesamoid bone of the 1st toe.
5. Base of the 1st metatarsal.

Comments:
Metatarsals are composed a of widened **base** proximally, followed by a **shaft**, **neck** and **head**. The **proximal**, **middle** and **distal phalanges** of the 2nd–5th toes, and the proximal and distal phalanges of the 1st toe also have a base, shaft and head.

Exam tip:
- This foot radiograph also displays a very common anatomical variant— fusion of the middle and distal phalanges of the 5th toe.

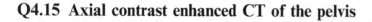

Q4.15 Axial contrast enhanced CT of the pelvis

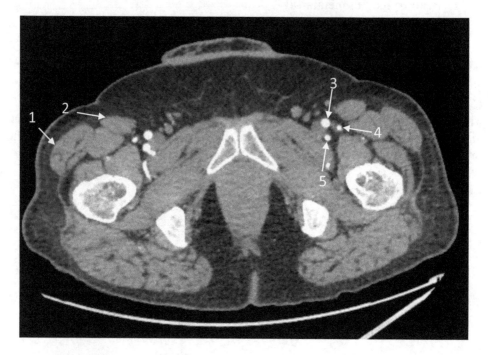

1. Name the arrowed structure.
2. Name the arrowed structure.
3. Name the arrowed structure.
4. Name the arrowed structure.
5. Name the arrowed structure.

Answers
1. Right tensor fascia lata.
2. Right sartorius.
3. Left superficial femoral artery.
4. Left lateral circumflex artery.
5. Left profunda femoris artery.

Comments:

The **sartorius muscle** attaches to the **anterior superior iliac spine**, and to the **medial tibia** as one of the constituent tendons of the pes anserinus.

The **lateral circumflex artery** is one of the first branches of the **profunda femoris artery**, after the **medial circumflex artery**. It can be identified because it passes anterior to the **femoral neck**, whilst the medial circumflex passes posterior to it. On the above image, it is more lateral and anterior than the medial circumflex (which is not labelled).

Exam tip:
- Identifying the level of the axial slice will help you correctly identify all the vascular structures. In the example above you can identify **symphysis pubis** and **greater trochanters** of the femurs. This indicates that we are below the level of the inguinal ligament and therefore in the territory of the femoral artery and its branches.

Q4.16 Axial CT angiogram of the lower limbs

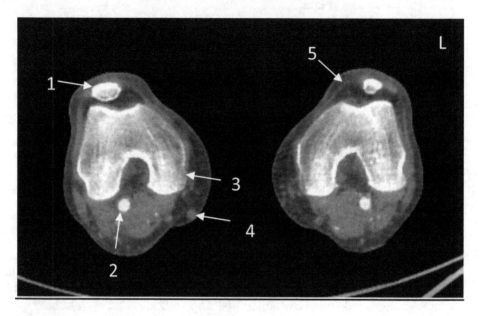

1. Name the arrowed structure.
2. Name the arrowed structure.
3. Name the arrowed structure.
4. Name the arrowed structure.
5. Name the arrowed structure.

Answers
1. Right patella.
2. Right popliteal artery.
3. Medial condyle of the right femur.
4. Right long saphenous vein.
5. Left medial patellofemoral retinaculum.

Comments:
The **popliteal artery** is the only major arterial structure passing behind the knee joint. The **long saphenous vein** can be differentiated from the **short saphenous vein** because it is on the medial aspect of the lower limb while the **short saphenous vein** is on the lateral side.

Exam tip:
- Recognising laterality is a key part of the exam. When there are two paired structures visible on the image, you will be expected to specify the laterality. As well as recognising the right side from the left, you will also have to distinguishing medial from lateral aspects of a structure—as with the medial and lateral condyles above.

Q4.17 Axial CT angiogram of the lower limbs

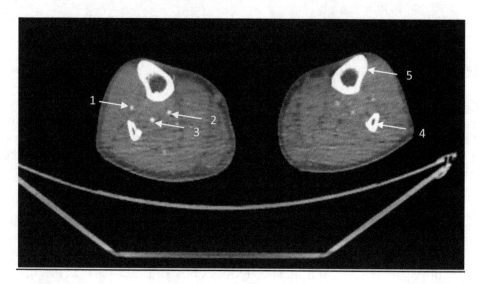

1. Name the arrowed structure.
2. Name the arrowed structure.
3. Name the arrowed structure.
4. Name the arrowed structure.
5. Name the arrowed structure.

Answers
1. Right anterior tibial artery.
2. Right posterior tibial artery.
3. Right fibular artery (peroneal artery).
4. Left fibula.
5. Left tibia.

Comments:
The **popliteal artery** trifurcates into three branches. First, it divides into the **anterior tibial artery** and the **tibioperoneal trunk**. The anterior tibial artery passes through the **interosseous membrane** between the **tibia** and **fibula**. It passes down through the anterior compartment and becomes the dorsalis pedis in the foot. The tibioperoneal trunk divides into the **posterior tibial artery** and the **fibular (peroneal) artery**, which is positioned more posteriorly and laterally.

Exam tip:
- The **anterior tibial artery** runs towards the anterior aspect of the tibia. The **posterior tibial artery** is located behind the tibia. Finally the **fibular artery** runs alongside the fibula.

Q4.18 Longitudinal image from a paediatric hip ultrasound

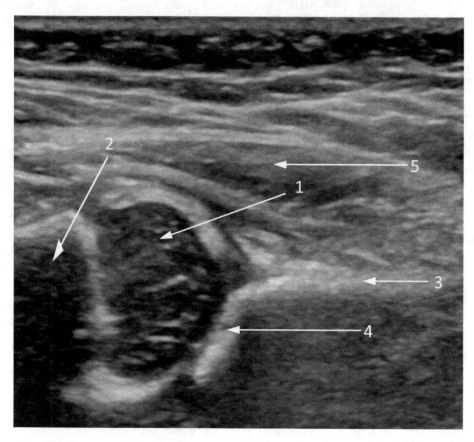

1. Name the arrowed structure.
2. Name the arrowed structure.
3. Name the arrowed structure.
4. Name the arrowed structure.
5. Name the arrowed structure.

Answers
1. Cartilage of the femoral head.
2. Proximal femoral metaphysis.
3. Iliac bone.
4. Bony acetabulum.
5. Gluteus muscles.

Comments:

This is a paediatric hip ultrasound. To orientate yourself, you must first realise that babies are placed on their side during a hip ultrasound. As such, the images are viewed horizontally.

As with any ultrasound, high density material (such as bone) appears highly echogenic. This can be seen at the iliac bone which is ossified at birth and appears as an almost linear, highly echogenic structure. The **proximal femur** and **femoral head** are not ossified at birth and are composed mainly of hyaline cartilage. As such they appear as low echogenicity on ultrasound scanning. The femoral head is oval in shape and centrally contains sinusoids which appear as echogenic 'worms', as you can see on the image above. The femoral head is enclosed in a capsule which covers the femoral head laterally.

The **acetabulum** articulates with the femoral head and can usually be seen as a rounded echogenic 'cup-like' structure which contains the femoral head.

As with adults, the gluteal muscles are large and powerful muscles which stabilise and move the hip joint. In this image above, it is difficult to delineate the three gluteal muscles (**gluteus maximus**, **gluteus medius** and **gluteus minimus**) and therefore '**gluteal muscles**' would suffice as an appropriate answer.

Some other important paediatric hip ultrasound anatomical landmarks that you should become familiar with are the **labrum**, the **triradiate cartilage**, the **greater trochanter** and the **acetabular cartilage**.

Exam tip:
- Remember to label anatomical landmarks as either bony or cartilaginous to obtain full marks.

Q4.19 Axial CT angiogram of the lower limbs

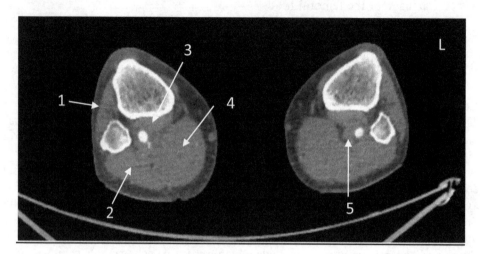

1. Name the arrowed structure.
2. Name the arrowed structure.
3. Name the arrowed structure.
4. Name the arrowed structure.
5. Name the arrowed structure.

Answers
1. Right tibialis anterior muscle.
2. Lateral head of the right gastrocnemius muscle.
3. Right popliteus muscle.
4. Medial head of the right gastrocnemius muscle.
5. Left popliteal vein.

Comments:
The **tibialis anterior muscle** runs along the anterior aspect of the tibia. The **gastrocnemius muscle** is divided into **lateral** and **medial heads**, which arise from the **lateral** and **medial femoral condyles** respectively. Gastrocnemius gives fibres to the **Achilles (calcaneal) tendon** which inserts onto the **tibial tuberosity**. Deep to gastrocnemius runs the **soleus muscle**.

Exam tip:
- If you are not sure what vascular structure the arrow is pointing to look out for contrast. The image above is labelled as a CT angiogram which implies that there will be contrast in the arteries. In the arterial phase, the veins should not be opacified.

Q4.20 Axial CT angiogram of the lower limbs

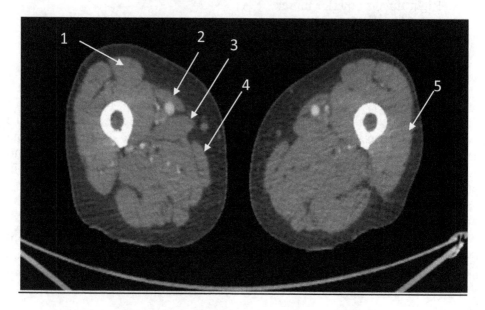

1. Name the arrowed structure.
2. Name the arrowed structure.
3. Name the arrowed structure.
4. Name the arrowed structure.
5. Name the arrowed structure.

Answers
1. Right rectus femoris muscle.
2. Right sartorius muscle.
3. Right adductor longus muscle.
4. Right gracillis muscle.
5. Left vastus lateralis muscle.

Comments:
There are three vasti muscles—**medialis, intermedius** and **lateralis**—which are located in the anterior compartment of the thigh. Two adductor muscles, **longus and brevis,** lie medially.

The **sartorius muscle** spirals from the lateral aspect of the thigh superiorly to the medial aspect inferiorly. It's tendon, together with the tendons of the **gracilis** and **semitendinous** muscles, forms the **pes anserius**, which inserts into the medial border of the proximal tibia.

Exam tip:
- Lower limb muscle anatomy is complex with multiple muscles crossing over and changing appearance depending on the level you are looking at. It is worth opening a lower limb CT scan and scrolling through the images in all three planes to get familiar with the structures and how they inter-relate as you move up and down the leg. Images in the exam are most likely to be at or near a joint, because it can be difficult to identify individual muscles in the mid-thigh or mid-calf on a single image.

Q4.21 Sagittal section from a CT of the cervical spine

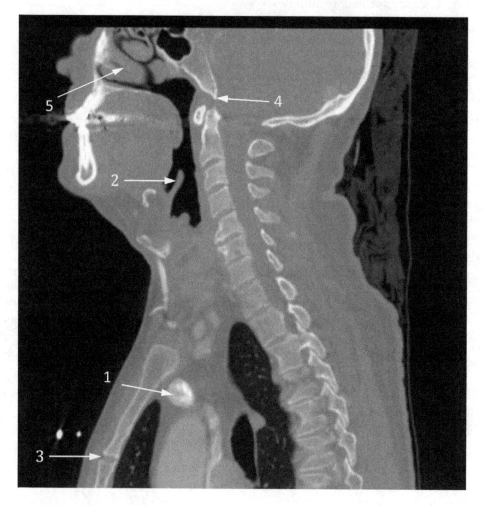

1. Name the arrowed structure.
2. Name the arrowed structure.
3. Name the arrowed structure.
4. Name the arrowed structure.
5. Name the arrowed structure.

Answers
1. Left brachiocephalic vein.
2. Epiglottis.
3. Manubriosternal joint.
4. Basion of the clivus.
5. Inferior nasal turbinate.

Comments:
The **epiglottis** is a leaf-shaped, flexible fibrocartilaginous structure that divides the **hypopharynx** from the **larynx**. The epiglottis projects postero-superiorly from its stem-like base. Its role is to protect the larynx during swallowing.

The **clivus** (Latin for 'slope') is a backwards-sloping midline structure formed by parts of the **occipital** and **sphenoid bones**. The tip of the clivus is known as the **basion,** and is in close proximity to the **odontoid peg** and the **anterior arch of C1**. The C1 vertebrae (commonly referred to as the atlas) is an atypical cervical vertebra with distinctive features. The atlas lacks a vertebral body and instead consists of a **posterior** and **anterior arch** that encircle the spinal cord. The anterior arch of C1 is connected to the odontoid peg by the **transverse ligament**.

The **manubriosternal joint** is the articulation of the inferior border of the **manubrium** and the superior border of the **sternal body**. It forms the clinical landmark of the sternal angle, and the **second costal cartilages** also articulate with the sternum at this point.

Exam tip:
- You may be familiar with seeing vascular structures in the axial plane, but you should also learn to identify them in coronal and sagittal slices.

Q4.22 Sagittal T1-weighted section of an MRI of the elbow

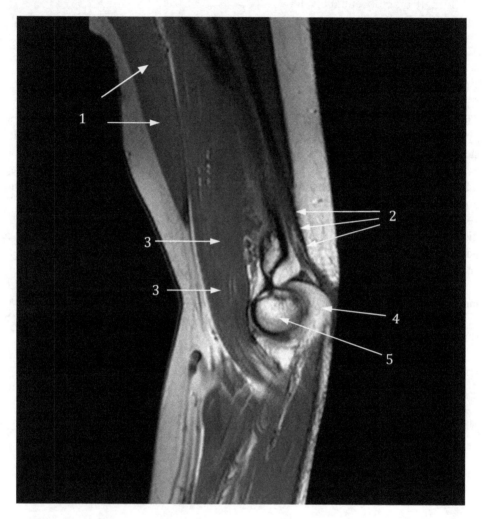

1. Name the arrowed structure.
2. Name the arrowed structure.
3. Name the arrowed structure.
4. Name the arrowed structure.
5. Name the arrowed structure.

Answers

1. Biceps brachii muscle.
2. Triceps brachii tendon.
3. Brachialis muscle.
4. Olecranon.
5. Trochlear of the humerus.

Comments:

The anterior (flexor) compartment of the upper arm contains the **biceps brachii**, coracobrachialis and **brachialis** muscles. Their blood supply is from the **brachial artery**, which is a branch of the **axillary artery**. The musculocutaneous nerve innervates muscles of this compartment.

Biceps brachii is formed from two heads—a **short head** and a **long head**. These two heads share a common single insertion onto the **radial tuberosity**. The muscle flexes the elbow joint and supinates the forearm.

The **brachialis** muscle lies deep to the biceps brachii on the anterior aspect of the humerus and is the main flexor of the elbow. It arises from the lower half of the humerus and inserts into the **coronoid process of the ulna**.

The posterior (extensor) compartment of the upper arms contains the **triceps brachii** and **anconeus** muscle. Blood supply is through the profunda brachii and the ulnar collateral arteries. The radial nerve innervates the muscles in this compartment. Triceps is formed of three heads—a lateral, a medial and a middle head. Distally, the three heads combine to share a common tendon insertion onto the postero-superior aspect of the **olecranon**.

Exam tip:

- To gain yourself valuable marks, be sure to label whether a structure is a muscle or a tendon. On MRI, tendons are dark linear structures that connect muscle to bone.

Q4.23 Axial fat-suppressed STIR sequence from a MRI of the wrist

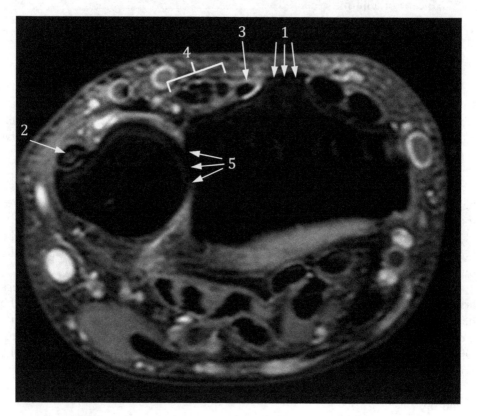

1. Name the arrowed structure.
2. Name the arrowed structure.
3. Name the arrowed structure.
4. Name the tendons which are in this compartment.
5. Name the arrowed structure.

Answers
1. Listers tubercle.
2. Extensor carpi ulnaris tendon.
3. Extensor pollicis longus tendon.
4. Extensor digitorum and extensor indices tendons.
5. Distal radio-ulna joint.

Comments:
The extensor tendons at the level of the wrist are divided into six numbered compartments stating from the most radial aspect to the ulnar aspect.

Extensor carpi ulnaris is the only tendon within compartment six. The role of this muscle is to extend and adduct the wrist. On a wrist MRI, the ECU is the most lateral of tendons and is commonly located within an indentation of the ulna bone.

Extensor pollicis longus crosses the wrist close to the midline and turns towards the thumb using **Lister's tubercle** as a pulley. EPL extends the interphalangeal joint of thumb. EPL is identifiable on an axial MRI wrist as it is the tendon immediately to the ulnar side of Lister's tubercle.

Compartment four is a dorsal and midline compartment which contains the tendons of **extensor digitorum** and **extensor indicis**. The role of these tendons is to extend the metacarpophalangeal joints of the fingers and wrist. This compartment is identifiable as a collection of tendons immediately ulnar to compartment 3 (containing extensor pollicis longus).

Exam tip:
- Use Lister's tubercle, a prominent dorsal protuberance on the distal radius, to navigate wrist tendon anatomy.

Q4.24 Lateral radiograph of a cervical spine

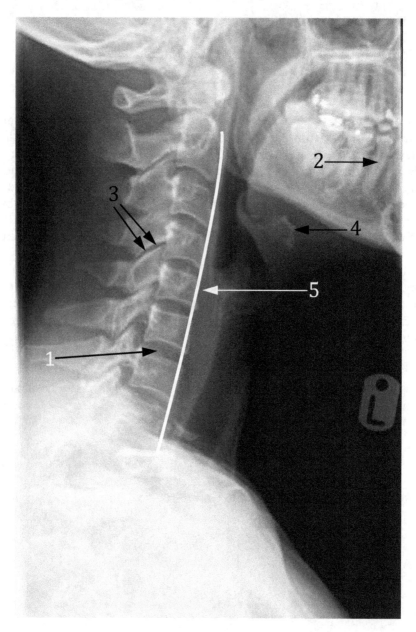

1. Name the arrowed structure.
2. Name the arrowed structure.
3. Name the arrowed structure.
4. Name the arrowed structure.
5. What is the name of this theoretical anatomical line?

Answers
1. C6/7 intervertebral disc.
2. Pulp chamber of a lower premolar.
3. C4/5 Facet joint.
4. Hyoid bone.
5. Anterior vertebral line.

Comments:
The cervical spine is made up of 7 vertebrae. The first two, C1 and C2, are given distinctive names based on their unique functions: **C1** is commonly referred to as the **atlas** and **C2** Is known as the **axis**. C3-C7 are more conventional cervical vertebrae. They comprise **vertebral bodies**, posterior to which are paired **pedicles, facet joints** and **laminae** and midline **spinous processes**.

Facet joints are articulations between two vertebrae. The **superior articular processes** connect with the corresponding **inferior articular processes** to form the joint.

The **anterior vertebral line** is a theoretical curved line that follows the lordotic curve of the cervical spine. The line should be smooth without any step-offs.

It is important to remember that teeth are also anatomical structures. Be sure to label whether the tooth is an incisor, canine, premolar or molar. You should also label if the tooth is situated on the upper jaw (**maxilla**) or lower jaw (**mandible**).

Exam tip:
- Hypothetical lines such as the anterior vertebral line, posterior vertebral line, spinolaminar line and the posterior spinous line can be identified on lateral c-spine imaging and could be asked in the exam.

Q4.25 Axial section from a CT of the chest

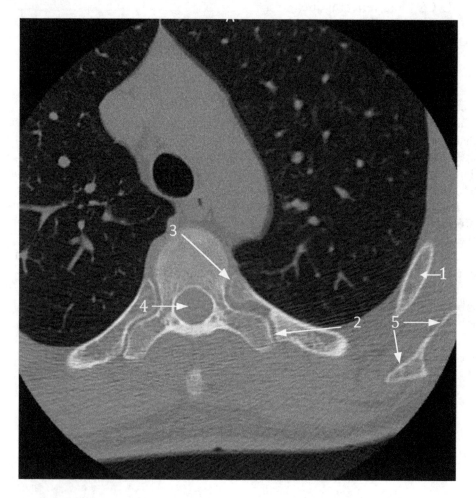

1. Name the arrowed structure.
2. Name the arrowed structure.
3. Name the arrowed structure.
4. Name the arrowed structure.
5. Name the arrowed structure.

Answers
1. Body of left rib.
2. Left costotransverse joint.
3. Left costovertebral joint.
4. Spinal canal.
5. Left scapula.

Comments:
The most anterior part of a thoracic vertebra is the **vertebral body**. The **pedicles** arise from the posterolateral margins of the vertebral body and extend posteriorly. The **laminae** extend infero-medially from the pedicles and fuse in the midline to form the base of the **spinous process**. The **transverse processes** arise at the junctions of the pedicles and laminae and extend laterally.

A thoracic vertebral body articulates with the **head of a rib** at the **costovertebral joint**. Each rib also articulates with a transverse process of the same vertebra at the **costotransverse joint**.

Exam tip:
- Try to be as precise as possible. You are more likely to obtain full marks if you give the specific part of the structure, such as the neck or body of a rib.

Q4.26 Axial STIR sequence from an MRI of the lumbar spine

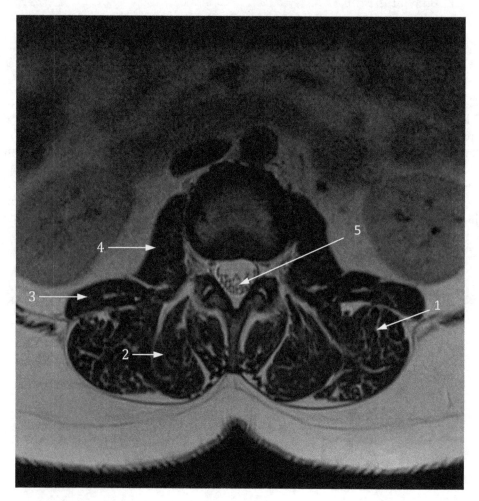

1. Name the arrowed structure.
2. Name the arrowed structure.
3. Name the arrowed structure.
4. Name the arrowed structure.
5. Name the arrowed structure.

Answers
1. Left Iliocostalis lumborum muscle.
2. Right longissmus muscle.
3. Right quadratus lumborum muscle.
4. Right psoas muscle.
5. Cauda equina.

Comments:
The presence of the kidneys and **cauda equina** in this image should help you to establish the approximate spinal level. The spinal cord usually ends at L1/2, and therefore this slice must be at the level of the lumbar vertebrae.

The erector spinae muscles are long, thin, vertical intrinsic muscles of the back. The group consists of **iliocostalis**, **longissimus** and **spinalis** (Useful mnemonic: I Love Standing).

Each muscle group of the erector spinae muscles can be further subdivided based on the anatomical position in which they are found (cervical, thoracic and lumbar). As a group, the erector spinae muscles are powerful extensors of the back.

The **psoas major muscle** and psoas minor muscle (if it is present), together with the **quadratus lumborum** muscle, lie in the paravertebral gutter and form the posterior abdominal wall.

Q4.27 Axial T1-weighted section from an MRI of the elbow

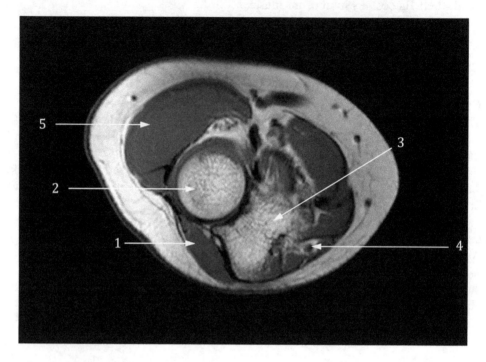

1. Name the arrowed structure.
2. Name the arrowed structure.
3. Name the arrowed structure.
4. Name the arrowed structure.
5. Name the arrowed structure.

Answers
1. Anconeus muscle.
2. Head of the radius.
3. Proximal ulna.
4. Ulna nerve.
5. Brachioradialis muscle.

Comments:
The best way to orientate yourself to this image is to identify the **radial head**. Once you have visualised this, the rest of the anatomy in this image should become clearer. The radial head is circular and articulates superiorly with the **capitellum of the humerus**. Medially it articulates with the **radial notch of the ulna**.

The small muscle on the posterior aspect of the elbow is the **anconeus**. It originates from the **lateral epicondyle** of the humerus and inserts onto the lateral aspect of the **olecranon**. The role of this muscle is to act as an extensor of the forearm.

The **brachioradialis** is a large muscle within the radial aspect of the forearm. It is the most superficial muscle within this region. The brachioradialis muscle inserts distally into the **radial styloid process**.

At the level of the elbow, **the ulna nerve** is positioned immediately posterior to the **medial epicondyle** of the humerus. It enters the forearm between the two heads of the flexor carpi ulnaris (humeral head and ulnar head).

Exam tip:
- The anconeus muscle is helpful in distinguishing the medial and lateral sides of the elbow on a single slice at the level of the olecranon. The anconeus lies laterally, and there are no comparable muscles on the medial side.

Q4.28 Axial T1-weighted section from an MRI of the shoulder

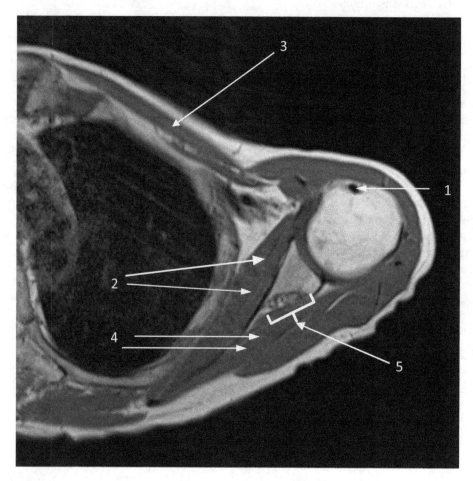

1. Name the arrowed structure.
2. Name the arrowed structure.
3. Name the arrowed structure.
4. Name the arrowed structure.
5. Name the nerve that runs through this anatomical space.

Answers
1. Left biceps tendon (long head).
2. Left subscapularis muscle.
3. Left pectoralis major muscle.
4. Left infraspinatus muscle.
5. Left suprascapular nerve.

Comments:
The stabilising muscles of the shoulder are called the rotator cuff muscles. They are the **supraspinatus, infraspinatus, teres minor** and **subscapularis muscles** (mnemonic: SITS).

The supraspinatus muscle originates from the supraspinous fossa on the posterior surface of the **scapula**, passes above the glenohumeral joint and inserts into the **greater tuberosity of the humerus**. It is above the axial level of this image.

The infraspinatus muscle originates from the infraspinous fossa of the scapula. It passes posteriorly across the glenoid and humeral head to insert on the posterior aspect of the greater tubercle of the humerus. It inserts immediately below the insertion of the supraspinatus.

Teres minor an elongated muscle which originates from the dorsal surface of the axillary border of the scapula. The muscle runs obliquely and laterally upwards. It inserts onto the inferior facet of greater tubercle of the humerus. This muscle is quite hard to identify on an axial image. You will most likely encounter this structure on a coronal image.

Subscapularis originates from the subscapular fossa of the anterior scapula and inserts into the **lesser tubercle** of the humerus.

The scapular notch is located at the supero-lateral aspect of the scapula. It is semi-circular in appearance and it is partly formed by the base of the **coracoid process**. The scapular notch is converted into a foramen by the superior transverse ligament. The suprascapular nerve runs within the notch and can be identified on MRI.

Exam tip:
- The rotator cuff anatomy takes a bit of time to get your head around- familiarise yourself with appearances on axial, coronal and sagittal views.

Q4.29 Axial plain radiograph of a shoulder

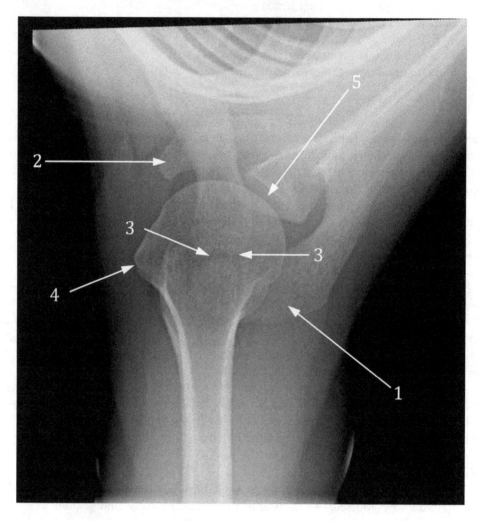

1. Name the arrowed structure.
2. Name the arrowed structure.
3. Name the arrowed structure.
4. Name a muscle that attaches to this anatomical structure.
5. Name the arrowed structure.

Answers
1. Acromion.
2. Coracoid process.
3. Acromioclavicular joint.
4. Supraspinatus, infraspinatus or teres minor muscles.
5. Glenoid fossa.

Comments:
The **head of the humerus** articulates with the **glenoid fossa** of the scapula. The **glenoid labrum** is a fibrocartilaginous structure that surrounds the glenoid and deepens the glenoid fossa.

The **acromioclavicular joint** is the articulation between the lateral end of the **clavicle** and the medial portion of the **acromion**. This joint is strengthened by the superior and inferior acromioclavicular ligaments.

Exam tip:
- The glenohumeral joint on a plain film radiograph can be viewed in may orientations. The view in this question is an axial view however be sure to familiarise yourself with the AP, Y-view and modified views of the shoulder. The hook-like appearance of the coracoid process can help you to identify the anterior side of the radiograph.

Q4.30 Coronal plain radiograph of an elbow

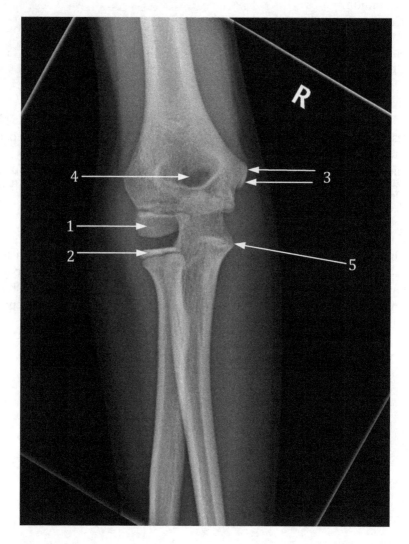

1. Name the arrowed structure.
2. Name the arrowed structure.
3. Name the arrowed structure.
4. Name the arrowed structure.
5. Name the arrowed structure.

Answers
1. Right capitellum epiphysis.
2. Right radial head apophysis.
3. Right internal epicondyle apophysis.
4. Right olecranon fossa.
5. Right coronoid process.

Comments:

The elbow has six distinct ossification centres that appear in a predictable chronological order. The mnemonic CRITOE will help you to remember when each ossification appears according the patients' age.
- Capitellum (age 1)
- Radial head (age 3)
- Internal epicondyle (age 5)
- Trochlea (age 7)
- Olecranon (age 9)
- External epicondyle (age 11)

This is a radiograph of a 10-year-old child. The external epicondyle epiphysis has not begun to ossify and is not visible.

Exam tip:
- The capitellum contributes to the growth of the humerus and is therefore considered an epiphysis. The other ossifications centres are apophyses.

Q4.31 Frontal radiograph of the shoulder

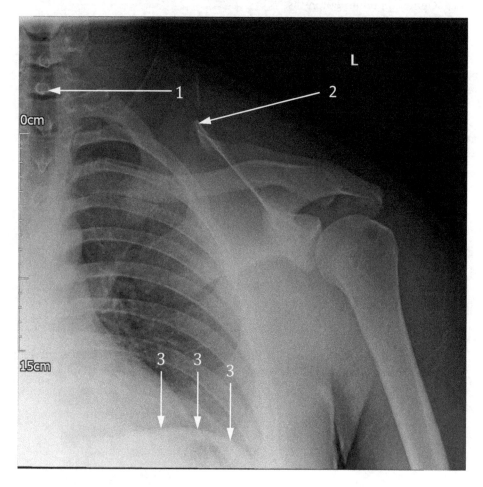

1. Name the arrowed structure.
2. Name the arrowed structure.
3. Name the arrowed structure.
4. Which fissure separates the right upper lobe and the middle lobe?
5. Name the anatomical variant.

Answers
1. C7 spinous process.
2. Superior angle of the left scapula.
3. Left hemidiaphragm.
4. The horizontal fissure.
5. Os acromiale.

Comments:

An 'os acromiale' is a failure of one of the three acromial ossification centres to fuse with the **acromion process**. It is therefore an anatomical variant and should not be mistaken for a fracture. Note the smooth outline and close association with the acromion.

There are three sub-types of os acromiale based on the location of non-union. A 'pre-acromion' os acromiale is seen in front of the acromioclavicular joint; a 'mesoacromion' is found behind the acromion and a meta-acromion is seen at the base of the acromion. On a single view it can be difficult to distinguish the subtypes and therefore simply stating 'Os acromiale' on this image would be sufficient to obtain the full marks.

Lung fissures separate the lung into lobes, and are good candidates to include in an exam. The **right oblique fissure** is slightly complicated as the superior part of the fissure separates the right upper lobe from the right lower lobe and the inferior part separates the middle lobe from the right lower lobe. The horizontal fissure is a unilateral structure which separates the **right upper lobe** and the **middle lobe**. The **left oblique fissure** separates **the left upper lobe** and the **left lower lobe.**

Exam tip:
- Strictly speaking, question 4 would not be asked in the examination, as there is no associated arrow. However, the anatomical knowledge underpinning it could be tested.

Q4.32 Longitudinal view from an ultrasound of a paediatric lumbar spine

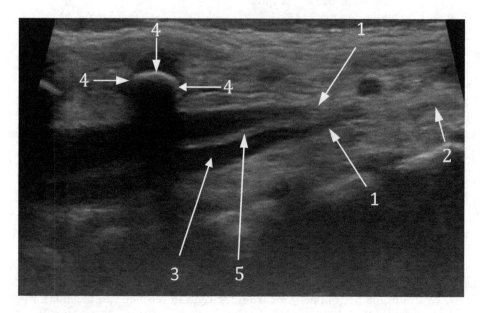

1. Name the arrowed structure.
2. Name the arrowed structure.
3. Name the arrowed structure.
4. Name the arrowed structure.
5. Name the arrowed structure.

Answers

1. Conus medularis.
2. Filum terminale.
3. Spinal cord.
4. Spinous process.
5. Central echo complex.

Comments:

At birth, the tip of the **conus medullaris** is situated roughly at the L2/3 vertebral body, whereas by 3 months is can been seen at the level of L1/2. On ultrasound the conus medularis is seen as the tapered tip of the **spinal cord**.

The **filum terminale** is a band of tissue that extends inferiorly from the conus medullaris. The role of the filum terminale is to provide longitudinal support to the spinal cord and help to anchor it in place. On ultrasound, the filum terminale is seen as a hyperechoic cord extending inferiorly from the conus medularis.

The spinal cord appears a tubular hypoechoic on structure with hyperechoic walls. The subarachnoid space surrounding the cord is hypoechoic.

The hyperechoic structure running through the middle of the spinal cord is referred to as the **central echo complex** and is thought to represent the central canal of the spinal cord.

Q4.33 Upper limb and cervical spine arterial angiogram

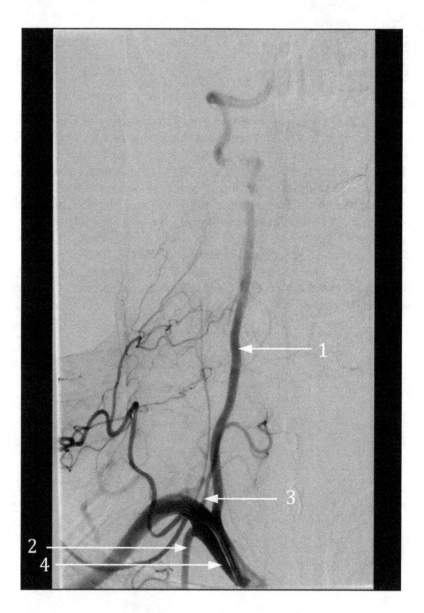

1. Name the arrowed structure.
2. Name the arrowed structure.
3. Name the arrowed structure.
4. Name the arrowed structure.
5. At which anatomical landmark does the subclavian artery become the axillary artery?

Answers
1. Vertebral artery.
2. Internal thoracic artery.
3. Thyro-cervical trunk.
4. Subclavian artery.
5. Lateral border of the 1st rib.

Comments:
On the right side of the body the subclavian artery is a branch of the **brachiocephalic trunk**. On the left, the subclavian artery arises directly from the **arch of the aorta**.

There are three parts of the subclavian artery (although this is beyond the level of detail needed for the exam). The first part is from its origin to the medial border of scalenus anterior; the second part is posterior to scalenus anterior; and the third part is from the lateral border of scalenus anterior to the lateral border of the first rib. Beyond the lateral border of the first rib, it becomes the **axillary artery**.

The first branch of the **subclavian artery** is the **vertebral artery**. This artery courses superiorly in the neck where it joins with the contralateral vertebral artery to form the **basilar artery** at the pontomedullary junction.

The **internal thoracic artery** arises from the subclavian artery opposite the vertebral artery and traverses caudally on either side of the sternum to supply the anterior chest.

The **thyrocervical trunk** is the third artery to arise from the subclavian artery, just lateral to the vertebral artery. The thyrocervical trunk supplies the neck, brachial plexus, and the scapular anastomosis. The next major branch is the **costocervical trunk**.

Exam tip:
- Although this question looks daunting, the best way to orientate yourself is to first identify the vertebral artery, with its distinctive appearance. The other arteries should be identifiable based on their relationship to the vertebral artery.

Q4.34 Transverse view from an ultrasound of a paediatric lumbar spine

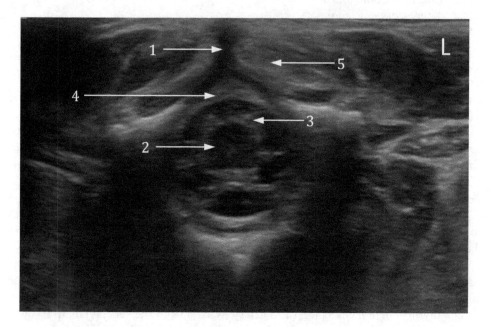

1. Name the arrowed structure.
2. Name the arrowed structure.
3. Name the arrowed structure.
4. Name the arrowed structure.
5. Name the arrowed structure.

Answers
1. Un-ossified spinous process of the lumbar spine.
2. Spinal cord.
3. Left dorsal nerve roots.
4. Posterior dura.
5. Left paravertebral muscles.

Comments:
This is an axial ultrasound of a lumbar vertebra in a neonate. During the neonatal period, ultrasound scans of this region can demonstrate great anatomical detail. The **spinous process** is seen as a hypoechoic structure and has a similar outline to adults. You can also appreciate the un-ossified **transverse process** in this image. The **spinal cord** is also hypoechoic but it contains a central hyperechoic midline structure, thought to represent the central spinal canal.

Dorsal nerve roots are paired structures which are slightly hyperechoic in echotexture.

The **dura** is a membrane of connective tissue which surrounds the spinal cord and is the outermost layer of the meninges. In adults the dura matter generally extends to the level of S3.

Exam tip:
- Un-ossified structures on ultrasound appear as hypoechoic structures. Be careful not to label un-ossified structures as bones.

Q4.35 Frontal radiograph—peg view

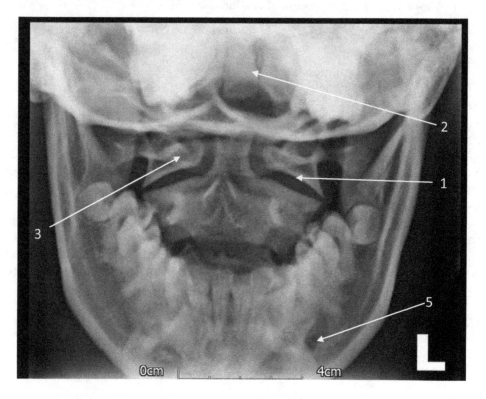

1. Name the arrowed structure.
2. Name the arrowed structure.
3. Name the arrowed structure.
4. Name one ligament which attach the axis (C2) to the occipital bone.
5. Name one structure which passes through the arrowed structure.

Answers
1. Inferior articular surface of the left lateral mass of C1.
2. Upper incisor.
3. Right lateral mass of C1.
4. Median apical ligament, alar ligaments, tectorial membrane or the longitudinal parts of the cruciform ligaments.
5. Left mental nerve (a branch of inferior alveolar nerve), left mental artery (a branch of inferior alveolar artery) or left mental vein (a tributary of inferior alveolar vein).

Comments:
This is an odontoid peg view, which provides important clinical and anatomical detail of the atlas (C1) and the axis (C2). The view is obtained with the patient opening their mouth as widely as possible. Teeth are commonly visible on this view.

The **atlas (C1)** is the first vertebra of the spine. It is formed by an **anterior arch, posterior arch, lateral masses** and **transverse process**. The articulations of the atlas (including the **atlanto-occipital joint**, median atlanto-axial joint and lateral atlanto-axial joint) allow flexion, extension, lateral flexion and rotation of the head.

The **axis (C2)** is the second vertebral body which is unique in structure—it contains the **odontoid peg (dens)** which is an embryological remnant of the primitive C1 vertebral body. In addition to the dens, the major components of the C2 vertebra includes the lateral masses, transverse processes, pedicles, laminas and **spinous process**.

The rounded lucent structure seen overlying the mandible is the **mental foramen**. It is an opening on the anterior aspect of the mandible and allows passage of the mental nerve, a branch of the inferior alveolar nerve which provides sensation to the anterior lower jaw and surrounding area. The mental foramen also transmits the mental artery and vein.

Exam tip:
- Remember to learn the contents of clinically relevant foramina, for example the mental foramen and the obturator foramen.

Q4.36 Coronal CT of the wrist

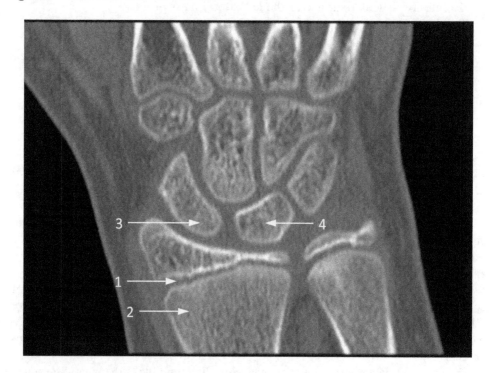

1. Name the arrowed structure.
2. Name the arrowed structure.
3. Name the arrowed structure.
4. Name the arrowed structure.
5. Name the intra-articular ligament which binds the structures in (3) and (4).

Answers
1. Distal radial physis.
2. Distal radial metaphysis.
3. Proximal pole of the scaphoid bone.
4. Lunate bone.
5. Scapho-lunate ligament.

Comments:

This is an unenhanced CT scan of a paediatric wrist. You can identify that this is a paediatric patient by the presence of an epiphysis and physeal plate. The age of closure of the distal radial and ulnar growth plate is variable, but is expected to happen by the time a person reaches their early twenties.

At birth, the wrist contains no calcified bones. The approximate order of wrist bone ossification is as follows:

Capitate: 1–3 months
Hamate: 2–4 months
Triquetrum: 2–3 years
Lunate: 2–4 years
Scaphoid: 4–6 years
Trapezium: 4–6 years
Trapezoid: 4–6 years
Pisiform: 8–12 years

The **scaphoid** is usually divided into a **distal pole**, a **waist** and a **proximal pole**. The blood supply to the scaphoid bone is via a branch of the radial artery. The proximal pole of the scaphoid is supplied by a retrograde vessel. The scaphoid articulates with the **radius**, the **trapezoid**, the **trapezium**, the **lunate** and the **capitate**.

The lunate is another bone within the proximal carpal row, which lies immediately medial to the scaphoid. It is named for its crescentic shape (from the Latin 'luna', meaning moon).

The **scapho-lunate ligament** is one of many ligaments that provides structural support to the proximal row of the carpal bones.

Exam tip:
- Remember that you only need to label left or right if there are two paired structures visible on the image provided.

Q4.37 Sagittal T2-weighted sequence from an MRI of the cervico-thoracic spine

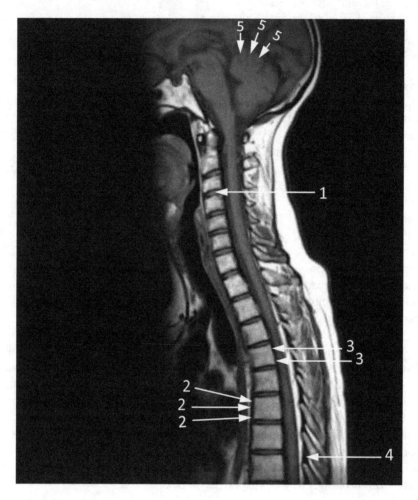

1. Name the arrowed structure.
2. Name the arrowed structure.
3. Name the arrowed structure.
4. Name the arrowed structure.
5. Name the arrowed structure.

Answers

1. Nucleus pulposus of C3/C4 intervertebral disc.
2. Anterior longitudinal ligament adjacent to the T6 vertebral body.
3. Posterior longitudinal ligament adjacent to the T4 vertebral body.
4. Ligamentum flavum at the level T7–T8.
5. Folia of the cerebellum.

Comments:

Intervertebral discs are composed of an outer **annulus fibrosus** and an inner **nucleus pulposus**. The nucleus pulposus is usually of higher signal intensity than the annulus fibrosus. The role of intervertebral discs is to protect the spinal cord and act as a shock-absorber for the spine. As you can see in this image, the upper thoracic intervertebral discs are usually thinner than cervical intervertebral discs.

The **anterior longitudinal ligament** traverses the anterior aspect of the vertebral bodies and is strongly attached to the periosteum of the bone and the anterior aspect of the annulus fibrosus of intervertebral discs. The **posterior longitudinal ligament** is loosely attached to the posterior aspect of the vertebral bodies, but strongly attached to the posterior annulus fibrosus. Each **ligamentum flava** is a thick ligament which connects the laminae of adjacent vertebrae. Ligaments on this image are demonstrated as low-signal, well-defined, linear structures.

CPSIA information can be obtained
at www.ICGtesting.com
Printed in the USA
BVHW062301260220
573455BV00004B/44